AF849705

Development of Agro-Industrial Wastes as Adsorbent Materials

This book examines advanced processing methodologies and innovative technologies for converting agricultural wastes and biomass into highly efficient biosorbent and adsorbent materials designed for environmental hazard remediation through cutting-edge and emerging technological approaches. Additionally, the book delivers comprehensive data regarding biosorbent and adsorbent material performance against diverse chemical contaminants, including synthetic dyes and heavy metals. Furthermore, it thoroughly analyzes future prospects, contemporary developments, and the commercial potential of biosorbent technologies.

Features:

- Focusses on the use of agricultural waste and biomass for the development of adsorbent materials.
- Gives updated information about the novel and emerging technologies for biosorbent preparation.
- Contains comprehensive information about the adsorption models.
- Highlights reduced use of toxic chemicals for biosorbent materials preparation.
- Fills the gap in environmental sustainability concept in the development of adsorbent materials.

This book is aimed at graduate students and researchers in agro-industrial, bioprocessing, environmental, and agricultural engineering.

Development of Agro-Industrial Wastes as Adsorbent Materials

Applications and Future Scope

Edited by
Sumit Kumar
Naveen Kumar Arora
Krushna Prasad Shadangi

CRC Press
Taylor & Francis Group
Boca Raton London New York

CRC Press is an imprint of the
Taylor & Francis Group, an **informa** business

Designed cover image: Shutterstock

First edition published 2026
by CRC Press
2385 NW Executive Center Drive, Suite 320, Boca Raton FL 33431

and by CRC Press
4 Park Square, Milton Park, Abingdon, Oxon, OX14 4RN

CRC Press is an imprint of Taylor & Francis Group, LLC

© 2026 selection and editorial matter, Sumit Kumar, Naveen Kumar Arora and Krushna Prasad Shadangi; individual chapters, the contributors

Reasonable efforts have been made to publish reliable data and information, but the author and publisher cannot assume responsibility for the validity of all materials or the consequences of their use. The authors and publishers have attempted to trace the copyright holders of all material reproduced in this publication and apologize to copyright holders if permission to publish in this form has not been obtained. If any copyright material has not been acknowledged please write and let us know so we may rectify in any future reprint.

Except as permitted under U.S. Copyright Law, no part of this book may be reprinted, reproduced, transmitted, or utilized in any form by any electronic, mechanical, or other means, now known or hereafter invented, including photocopying, microfilming, and recording, or in any information storage or retrieval system, without written permission from the publishers.

For permission to photocopy or use material electronically from this work, access www.copyright.com or contact the Copyright Clearance Center, Inc. (CCC), 222 Rosewood Drive, Danvers, MA 01923, 978-750-8400. For works that are not available on CCC please contact mpkbookspermissions@tandf.co.uk

For Product Safety Concerns and Information please contact our EU representative GPSR@taylorandfrancis.com. Taylor & Francis Verlag GmbH, Kaufingerstraße 24, 80331 München, Germany.

Trademark notice: Product or corporate names may be trademarks or registered trademarks and are used only for identification and explanation without intent to infringe.

ISBN: 978-1-032-68504-5 (hbk)
ISBN: 978-1-041-28940-1 (pbk)
ISBN: 978-1-003-76725-1 (ebk)

DOI: 10.1201/9781003767251

Typeset in Times
by SPi Technologies India Pvt Ltd (Straive)

Contents

About the Editors ... vii
List of Contributors ... viii
Preface ... xi
Acknowledgements ... xii

Chapter 1 Introduction ... 1

Sumit Kumar, Navin Chandra Shahi, and Giovana Ferrari

Chapter 2 Characteristics of Biomass and Agro-Industrial-Based Materials ... 8

Shwetha Papani and M. S. Shwetha

Chapter 3 Synthetic Dyes, Heavy Metals, and Other Hazardous Materials for the Environment ... 26

G. H. Kavana, S Divya, Kamalika Sarkar, and Sumit Kumar

Chapter 4 Adsorption Mechanism Demonstrating Adsorbent–Adsorbate Interactions for Emerging Contaminants in the Aqueous Media ... 41

Susmita Singha Roy, Asmita Mishra, and Nirmalendu Sekhar Mishra

Chapter 5 Adsorption Mechanism and Kinetics for Water Treatment ... 53

Mukul Swain and Krushna Prasad Shadangi

Chapter 6 Novel Technologies for Preparation of Adsorbent Materials ... 91

Tapas Roy, Sheetal N. Mane, Purnima Kumari, and Anupama Singh

Chapter 7 Advanced Oxidation Processes for Wastewater Treatment ... 114

Amit Kumar Behera and Krushna Prasad Shadangi

Chapter 8 Microstructure Characterization and Characterization Techniques of Adsorbent Materials ... 127

Diya Kuttappa, Sayonjit Roy, Rounak Ghosh, and Ishita Auddy

Chapter 9 Advantages and Disadvantages of Novel Technologies for Adsorbent from Agro-Based Industries 148

B Yashvashri and Sumit Kumar

Chapter 10 Batch and Continuous Biosorption Process and Its Scale-Up to Industrial Application 167

S Aishwaryalakshmi, S. K. Sivashree, Srinivasan Krishnamoorthy, R Jagan Mohan, N Baskaran, and S Vignesh

Chapter 11 Scope and Challenges 190

Pavel Rout, Soma Maji, and Sumit Kumar

Index 200

About the Editors

Sumit Kumar, PhD, is currently working as Assistant Professor, Department of Food Technology, JAIN (Deemed-to-be University), Bengaluru, Karnataka. He has completed his BTech in Agricultural Engineering and MTech in Process and Food Engineering from GB Pant University of Agriculture and Technology, Pantnagar, Uttarakhand, and his PhD in Food Process Engineering from the National Institute of Food Technology, Entrepreneurship and Management (NIFTEM), Thanjavur, Tamil Nadu. During his PhD, he has worked on the utilization of waste generated from the food processing industry and biomass for the development of adsorbent material and the extraction of valuable compounds. He has published 12+ National and International papers in peer-reviewed SCI journals and has four Indian Patents (published). Also, he is a reviewer for many national and international journals.

Naveen Kumar Arora, PhD (Microbiology), is Professor and Dean of the School of Earth and Environmental Sciences at Babasaheb Bhimrao Ambedkar University, Lucknow. He is a renowned researcher in the field of Environmental Microbiology and Biotechnology. His specific area of research is rhizosphere biology and plant growth-promoting rhizobacteria (PGPR). He has more than 100 research papers published in premium international journals. He is the editor of 45 books published by international publishers. He is a member of several national and international societies, serves on the editorial board of four journals, and is a reviewer of several international journals. As per the list released by Stanford University, USA, and in collaboration with Elsevier BV, he is among the top 2% scientists in the world. He is Editor-in-Chief of the journal *Environmental Sustainability* and Series Editor of the book series *Microorganisms for Sustainability*. He has been an advisor to 18 doctoral students. He has also received awards for excellence in research by the Honorable Governor of Uttar Pradesh, India, the Asian PGPR Society, and the Samagra Vikas Welfare Society. He is Fellow of the International Society for Environmental Botanists and has delivered talks as Keynote and Plenary Speaker in conferences around the world. He has collaborations with researchers in various countries, including Germany, Uzbekistan, the USA, Turkey, etc. Although an academician and researcher by profession, he has a huge obsession with wildlife and its conservation and has authored a book, *Splendid Wilds*. He is President of the Society for Conservation of Wildlife and is also Secretary of the Society for Environmental Sustainability.

Krushna Prasad Shadangi, PhD, is working as an assistant professor in the Department of Chemical Engineering at Veer Surendra Sai University of Technology, Burla, Odisha, India. He earned his doctorate in chemical engineering from the Indian Institute of Technology (IIT), Guwahati, India. He has ten years of research experience in the field of biofuel technologies and has contributed eight book chapters in edited books. He has published 30+ papers in peer-reviewed SCI journals and is an editorial board member of five international journals.

Contributors

S Aishwaryalakshmi
National Institute of Food Technology, Entrepreneurship and Management
Thanjavur, India

Ishita Auddy
Department of Food Technology
Faculty of Engineering and Technology
JAIN (Deemed-to-be University)
Bangalore, India

N Baskaran
National Institute of Food Technology, Entrepreneurship and Management
Thanjavur, India

Amit Kumar Behera
Department of Chemical Engineering
Veer Surendra Sai University of Technology
Burla, India

S Divya
Department of Food Technology
Faculty of Engineering and Technology
JAIN (Deemed-to-be University)
Bangalore, India

Giovana Ferrari
Department of Industrial Engineering
University of Salerno
Fisciano, Italy

Rounak Ghosh
Department of Food Technology
Faculty of Engineering and Technology
JAIN (Deemed-to-be University)
Bangalore, India

G. H. Kavana
Department of Food Technology
Faculty of Engineering and Technology
JAIN (Deemed-to-be University)
Bangalore, India

Srinivasan Krishnamoorthy
National Institute of Food Technology, Entrepreneurship and Management
Thanjavur, India

Sumit Kumar
Department of Food Technology
Faculty of Engineering and Technology
JAIN (Deemed-to-be University)
Bangalore, India

Purnima Kumari
Department of Food Engineering
National Institute of Food Technology, Entrepreneurship and Management
Kundli, India

Diya Kuttappa
Department of Food Technology
Faculty of Engineering and Technology
JAIN (Deemed-to-be University)
Bangalore, India

Soma Maji
Department of Food Technology
Faculty of Engineering and Technology
JAIN (Deemed-to-be University)
Bangalore, India

Sheetal N. Mane
Department of Food Engineering
National Institute of Food Technology, Entrepreneurship and Management
Kundli, India

Asmita Mishra
School of Bio Sciences and Technology
Vellore Institute of Technology
Vellore, India

Nirmalendu Sekhar Mishra
Department of Environmental Science and Engineering
SRM University AP
Amaravati, India

R Jagan Mohan
National Institute of Food Technology, Entrepreneurship and Management
Thanjavur, India

Shwetha Papani
Department of Food Technology
Faculty of Engineering and Technology
JAIN (Deemed-to-be University)
Bangalore, India

Pavel Rout
Department of Dairy Chemistry
College of Dairy Science and Food Technology
Raipur, India

Sayonjit Roy
Department of Food Technology
Faculty of Engineering and Technology
JAIN (Deemed-to-be University)
Bangalore, India

Susmita Singha Roy
Department of Environmental Science and Engineering
SRM University AP
Amaravati, India

Tapas Roy
Department of Food Engineering
National Institute of Food Technology, Entrepreneurship and Management
Kundli, India

Kamalika Sarkar
Department of Food Technology
Faculty of Engineering and Technology
JAIN (Deemed-to-be University)
Bangalore, India

Krushna Prasad Shadangi
Department of Chemical Engineering
Veer Surendra Sai University of Technology
Burla, India

Navin Chandra Shahi
Department of Agricultural Engineering
Mahamaya College of Agricultural Engineering and Technology (MCAET)
Ambedkar Nagar, India

M. S. Shwetha
Department of Food Technology
Faculty of Engineering and Technology
JAIN (Deemed-to-be University)
Bangalore, India

Anupama Singh
Department of Food Engineering
National Institute of Food Technology, Entrepreneurship and Management
Kundli, India

S. K. Sivashree
National Institute of Food Technology, Entrepreneurship and Management
Thanjavur, India

Mukul Swain
Graduate School of Advanced Science and Technology
Japan Advanced Institute of Science and Technology
Ishikawa, Japan

S Vignesh
National Institute of Food Technology, Entrepreneurship and Management
Thanjavur, India

B Yashvashri
Department of Food Technology
Faculty of Engineering and Technology
JAIN (Deemed-to-be University)
Bangalore, India

Preface

The adsorption of hazardous materials from aqueous solution, air, etc., is a well-known physicochemical process for the removal of contaminants. The dissolved contaminants lead to cancer in human beings after consumption. In the past 100 years, the adsorption processes have been used to a larger extent for the removal of contaminants from wastewater. On the contrary, agro-industrial-based adsorbent materials have been used for groundwater remediation or wastewater treatment. The present book will reveal the scope of the theoretical fundamentals of adsorption technology for wastewater treatment. In particular, it presents the most important basics for adsorbent materials prepared using agro-industrial waste, adsorption mechanism, process modeling, etc., for planning and evaluation of experimental adsorption studies. The book begins with an introduction to basic concepts and an overview of adsorption processes in water treatment, followed by a chapter on adsorbents and their characterization. The main chapters of the book deal with the three constituents of the practice-related adsorption theory: adsorption equilibria, adsorption kinetics, and adsorption dynamics in fixed-bed columns. Single-solute systems as well as multicomponent systems of known and unknown composition are considered. Special emphasis is given to the competitive adsorption of micropollutants and organic background compounds due to the high relevance for micropollutant removal from different types of water. The treatment of engineered processes ends with a chapter on the restoration of the adsorbent capacity by regeneration and reactivation. The contents of the book are completed by an outlook on geosorption processes, which play an important role in semi-natural treatment processes such as bank filtration or groundwater recharge. It was in the mid-1970s, at the beginning of my PhD studies, when I was first faced with the theme of adsorption. Although I have broadened my research field during my scientific career, adsorption has always remained in the focus of my interests. I would be pleased if this book, which is based on my long-term experience in the field of adsorption, would help readers to find easy access to the fundamentals of this important water treatment process. I would like to thank all those who contributed to this book by some means or other, in particular my PhD and MTech students, as well as numerous partners in different adsorption projects.

Sumit Kumar
Naveen Kumar Arora
Krushna Prasad Shadangi

Acknowledgements

The authors would like to thank senior publisher **Dr Gagandeep Singh** Sir of CRC Press for his consistent support throughout the completion of this work. Also, editors would like to thank all the chapter authors/contributors for their good work in making this book a great success. Editors would like to thank their family members and well-wishers for their continued support.

1 Introduction

Sumit Kumar, Navin Chandra Shahi, and Giovana Ferrari

1.1 INTRODUCTION

Agro-industrial activities generate substantial amounts of waste, including crop residues, food processing by-products, and biomass residues. Improper disposal of these wastes poses significant environmental and health risks, such as soil and water contamination, greenhouse gas emissions, and public health hazards (Iyer et al., 2024). However, these wastes also possess valuable properties that can be harnessed for various applications, including environmental remediation. Adsorption is a widely recognized method for the removal of pollutants from water and air due to its simplicity, cost-effectiveness, and efficiency. Adsorbent materials play a crucial role in this process by capturing and immobilizing contaminants onto their surfaces. Conventionally, activated carbon derived from coal, wood, or coconut shells has been the preferred choice for adsorption-based treatment systems (Tariq et al., 2024). However, the high cost and limited availability of activated carbon have prompted researchers to explore alternative adsorbents, including agro-industrial wastes. Agro-industrial wastes are abundant, renewable, and often underutilized resources that can be converted into effective adsorbent materials. By repurposing these wastes, not only can environmental pollution be mitigated, but the economic burden associated with waste management can also be reduced. This report aims to provide a comprehensive overview of the utilization of agro-industrial wastes as adsorbent materials for environmental remediation.

1.2 AGRO-INDUSTRIAL WASTES: SOURCES AND COMPOSITION

Agro-industrial wastes encompass a wide range of organic materials generated from agricultural, food processing, and forestry activities. These wastes can be broadly categorized into lignocellulosic biomass, agricultural residues, food processing by-products, and other organic wastes. Lignocellulosic biomass, such as crop residues (e.g., rice husk, wheat straw, and corn cob) and forestry residues (e.g., sawdust and wood chips), constitutes a significant portion of agro-industrial wastes (Sharma et al., 2024). These materials are rich in cellulose, hemicellulose, and lignin, which are the primary constituents of plant cell walls (Table 1.1). Agricultural residues, including crop stalks, leaves, and husks, are generated during harvesting and processing operations. These residues vary in composition depending on the type of crop and the processing methods employed. Food processing by-products, such as fruit peels, pomace, and bran, are generated during the production of food, beverages, and

DOI: 10.1201/9781003767251-1

TABLE 1.1
Agro-Industrial Waste Composition and Sources

Waste Material	Source Industry	Key Components	Typical Applications
Rice husk	Rice milling	Cellulose, lignin, silica	Adsorption of heavy metals, dyes
Sugarcane bagasse	Sugar industry	Cellulose, hemicellulose, lignin	Dye and metal ion adsorption
Peanut shells	Oil extraction/snack foods	Cellulose, lignin	Dye removal, metal adsorption
Coconut shells/fibers	Coconut processing	Lignocellulose, carbon content	Activated carbon production
Coffee husks	Coffee processing	Lignin, hemicellulose	Adsorption of dyes and metals
Fruit peels (banana, orange)	Juice/canning industry	Pectin, cellulose, phenolic compounds	Adsorption of heavy metals and organics
Wheat bran	Flour milling	Fiber, proteins, phenolics	Removal of organic pollutants
Palm kernel shells	Palm oil industry	Lignin, cellulose	Production of activated carbon

Source: Sharma et al. (2024), Chahal et al. (2024), Nyairo et al. (2025), and Okoro et al. (2025).

edible oils (Mubarak et al., 2024). These by-products often contain valuable nutrients and bioactive compounds, making them suitable candidates for various applications, including adsorbents (Tolkou et al., 2023). The composition of agro-industrial wastes can vary significantly based on factors such as crop type, geographic location, and processing techniques. However, they typically contain organic polymers, carbohydrates, proteins, lipids, lignin, and other biopolymers (Chahal et al., 2024). These organic constituents provide functional groups (e.g., hydroxyl, carboxyl, and amino) that facilitate adsorption interactions with pollutants. The details regarding the adsorption capacity are given in Section 1.2.

1.3 ADSORPTION MECHANISMS

Adsorption is a surface phenomenon whereby molecules or ions from a fluid phase adhere to the surface of a solid adsorbent. The adsorption process is governed by various mechanisms, including physical adsorption (physisorption) and chemical adsorption (chemisorption). Physical adsorption involves weak van der Waals forces or electrostatic interactions between the adsorbate molecules and the adsorbent surface. In contrast, chemical adsorption entails the formation of covalent or ionic bonds between the adsorbate and the adsorbent surface (Tolkou et al., 2024). The adsorption mechanisms involved in the removal of pollutants using agro-industrial adsorbents depend on several factors, including the nature of the adsorbate, the surface properties of the adsorbent, and the prevailing environmental conditions. Common mechanisms include electrostatic interactions, hydrogen bonding, π–π interactions, ion exchange, and surface complexation. The presence of functional groups (e.g., hydroxyl, carboxyl, and amino) on the adsorbent surface plays a crucial role in facilitating these interactions (Jahin et al., 2024). Factors influencing adsorption efficiency include the surface area, pore structure, surface chemistry, and surface charge of the adsorbent. Agro-industrial adsorbents with high surface area and abundant surface functional groups exhibit enhanced adsorption capacity and efficiency. Additionally, parameters such as pH, temperature, initial concentration of pollutants, contact time, and agitation speed can significantly influence the adsorption process (Duwiejuah et al., 2024).

1.4 AGRO-INDUSTRIAL WASTE-BASED ADSORBENTS

Agro-industrial wastes can be converted into adsorbent materials through various preparation methods, including physical, chemical, and biological techniques. Physical methods involve size reduction, drying, grinding, and thermal treatment of the raw material to enhance its surface properties. Chemical methods utilize chemical agents (e.g., acids, bases, and oxidizing agents) to modify the surface chemistry of the adsorbent and introduce functional groups that enhance adsorption capacity (Figure 1.1). Biological methods, such as microbial fermentation and enzymatic treatment, utilize biological agents to modify the structure and composition of agro-industrial wastes. These methods offer eco-friendly and sustainable approaches to adsorbent synthesis, utilizing microbial enzymes or metabolic pathways to enhance the adsorption properties of the materials (Kumar et al., 2024). Modification techniques, which include impregnation, activation, and functionalization, can further

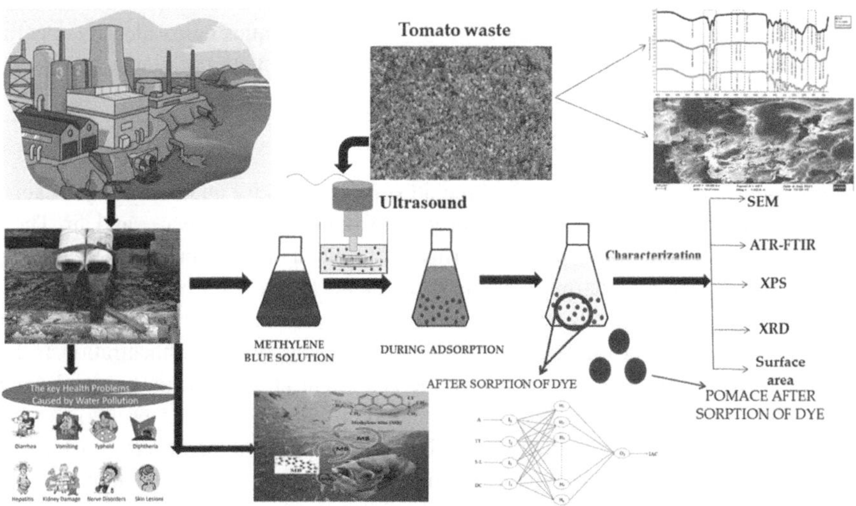

FIGURE 1.1 Tomato waste-based biosorbent material for the adsorption of synthetic dyes (Kumar et al., 2022b).

improve the adsorption performance of agro-industrial adsorbents. Impregnation involves the deposition of active agents (e.g., metal oxides and nanoparticles) onto the surface of the adsorbent to enhance its adsorption capacity. Activation methods, such as physical activation (e.g., carbonization and pyrolysis) and chemical activation (e.g., activation with acids and bases), increase the surface area and porosity of the adsorbent, thereby enhancing its adsorption efficiency. Functionalization techniques introduce specific functional groups onto the adsorbent surface to enhance its affinity for target pollutants (Alvez-Tovar et al., 2025).

1.5 ADSORBENT CHARACTERIZATION TECHNIQUES

The characterization of agro-industrial adsorbents is essential for understanding their physicochemical properties and adsorption behavior. Various analytical techniques can be employed to assess the surface morphology, surface area, pore structure, surface chemistry, and functional groups of the adsorbent. Common characterization techniques include scanning electron microscopy (SEM), transmission electron microscopy (TEM), Fourier-transform infrared spectroscopy (FTIR), X-ray diffraction (XRD), Brunauer–Emmett–Teller (BET) surface area analysis, and pore size distribution analysis. SEM and TEM provide detailed information about the surface morphology and microstructure of the adsorbent, allowing visualization of pore structures, particle size, and surface irregularities. FTIR spectroscopy is used to identify functional groups present on the adsorbent surface, providing insights into the chemical composition and surface chemistry of the material. XRD analysis enables the identification of crystalline phases and structural properties of the adsorbent (Kumar et al., 2022a). BET surface area analysis and pore size distribution analysis are crucial for determining the surface area, pore volume, and pore size distribution of the adsorbent. These parameters influence the adsorption capacity and efficiency of the material by providing access to active sites for pollutant adsorption. Functional group analysis using FTIR or other spectroscopic techniques helps identify specific functional groups responsible for adsorption interactions with pollutants.

1.6 ADSORPTION MODELS

Numerous studies have investigated the adsorption capacity and efficiency of agro-industrial adsorbents for the removal of various pollutants from water and air. These studies employ batch adsorption experiments to evaluate adsorption kinetics, equilibrium isotherms, and thermodynamic parameters. The adsorption kinetics describe the rate at which pollutants are adsorbed onto the adsorbent surface, while equilibrium isotherms depict the relationship between the adsorbate concentration in the solution and the amount adsorbed onto the adsorbent at equilibrium (Ahmad Aftab et al., 2024). The kinetic models, which include pseudo-first-order, pseudo-second-order, and intra-particle diffusion models, are used to analyze adsorption kinetics data and determine the rate-controlling mechanisms. Equilibrium isotherm models, including Langmuir, Freundlich, and Temkin isotherms, are employed to evaluate the adsorption capacity and affinity of the adsorbent for the adsorbate. Thermodynamic parameters, such as Gibbs free energy (ΔG°), enthalpy (ΔH°), and entropy (ΔS°), provide insights into the spontaneity, exothermicity, and randomness of the adsorption process (Kumar et al., 2022b). Several studies on the removal of pollutants using agro-industrial adsorbents encompass a wide range of contaminants, including heavy metals, dyes, pesticides, pharmaceuticals, and emerging pollutants. These studies demonstrate the effectiveness of agro-industrial adsorbents in removing pollutants from aqueous solutions and air streams, highlighting their potential for environmental remediation applications (Kumar et al., 2022a). Comparisons with conventional adsorbents, such as activated carbon and synthetic resins, reveal the competitive performance of agro-industrial adsorbents in terms of adsorption capacity, cost-effectiveness, and sustainability.

1.7 APPLICATION OF ADSORBENTS

Agro-industrial adsorbents find diverse applications in environmental remediation, including water treatment, wastewater treatment, and air purification. In water treatment applications, these adsorbents are used for the removal of heavy metals, organic pollutants, nutrients, and microbial contaminants from surface water, groundwater, and industrial effluents. The abundant surface functional groups and high adsorption capacities of agro-industrial adsorbents make them effective for the removal of a wide range of contaminants, including heavy metals (e.g., lead, cadmium, and arsenic), dyes (e.g., methylene blue and Congo red), pesticides (e.g., atrazine and paraquat), pharmaceuticals (e.g., antibiotics and hormones), and emerging pollutants (e.g., per- and polyfluoroalkyl substances and pharmaceuticals and personal care products (Bekchanov et al., 2024).

In wastewater treatment applications, agro-industrial adsorbents are employed for the removal of pollutants from various industrial effluents, including textile, tannery, pulp and paper, pharmaceutical, and chemical industries. These adsorbents offer an eco-friendly and sustainable alternative to conventional treatment methods, such as chemical precipitation, coagulation–flocculation, and membrane filtration (Kumar et al., 2022b). The versatility and cost-effectiveness of agro-industrial adsorbents make them suitable for large-scale applications in wastewater treatment plants, where they can help meet regulatory standards and minimize environmental impacts. In air purification applications, agro-industrial adsorbents are utilized for the removal of gaseous pollutants, including volatile organic compounds (VOCs), odors, and hazardous gases, from indoor and outdoor air environments. These adsorbents can be incorporated into air filtration systems, adsorption columns, or packed bed reactors to capture and immobilize airborne contaminants effectively (Verma et al., 2024). The porous structure and high surface area of agro-industrial adsorbents enable efficient removal of gaseous pollutants through physical adsorption mechanisms, reducing air pollution and improving indoor air quality. Emerging applications of agro-industrial adsorbents include the removal of emerging pollutants, such as microplastics, pharmaceuticals, and personal care products, from water and wastewater streams. These pollutants pose significant challenges to environmental and human health due to their persistence, bioaccumulation potential, and adverse effects on aquatic ecosystems. Agro-industrial adsorbents offer a promising solution for mitigating the environmental impact of emerging pollutants by providing cost-effective and sustainable adsorption materials.

1.8 CHALLENGES AND FUTURE SCOPE

Despite their numerous advantages, the utilization of agro-industrial adsorbents faces several challenges that need to be addressed to realize their full potential for environmental remediation. These challenges include variability in adsorption performance due to differences in feedstock composition, limited availability of standardized characterization methods, and potential leaching of contaminants from the adsorbent into the environment. Additionally, scalability issues, economic feasibility, and regulatory considerations may affect the practical implementation of agro-industrial adsorbents on a large scale. To overcome these challenges, future research efforts should focus on optimizing adsorbent synthesis methods, improving adsorption performance through material modification and functionalization, developing standardized characterization protocols, and conducting rigorous long-term studies to assess the environmental impact and sustainability of agro-industrial adsorbents. Collaboration between academia, industry, and government agencies is essential to facilitate technology transfer, promote innovation, and address regulatory barriers to the adoption of agro-industrial adsorbents in real-world applications.

1.9 CONCLUSION

Agro-industrial wastes represent a valuable and underutilized resource that can be converted into effective adsorbent materials for environmental remediation. By repurposing these wastes, not only can environmental pollution be mitigated, but also the economic burden associated with waste management can be reduced. Agro-industrial adsorbents offer numerous advantages, including abundance, renewability, cost-effectiveness, and sustainability, making them attractive alternatives to conventional adsorbents such as activated carbon. Through a combination of physicochemical properties, surface functional groups, and pore structures, agro-industrial adsorbents exhibit high adsorption capacities and efficiencies for a wide range of pollutants in water and air. These adsorbents find diverse applications in water treatment, wastewater treatment, air purification, and emerging pollutant removal, offering eco-friendly and sustainable solutions to environmental challenges. However, several challenges need to be addressed to realize the full potential of agro-industrial adsorbents, including variability in adsorption performance, scalability issues, economic feasibility, and regulatory considerations. Future research efforts should focus on optimizing adsorbent synthesis methods, improving adsorption performance through material modification, developing standardized characterization protocols, and addressing regulatory barriers to adoption. In conclusion, the utilization of agro-industrial wastes as adsorbent materials represents a promising avenue for sustainable environmental remediation, offering a synergistic approach to waste management and pollution control.

REFERENCES

Ahmad Aftab, R., Yusuf, M., Ahmad, F., Danish, M., Zaidi, S., Vo, D. V. N., & Ibrahim, H. (2024). Green technology approach towards the removal of heavy metals, dyes, and phenols from water using agro-based adsorbents: A review. *Chemistry – An Asian Journal*, *19*(19), e202400154.

Alvez-Tovar, B., Scalize, P. S., Angiolillo-Rodríguez, G., Albuquerque, A., Ebang, M. N., & de Oliveira, T. F. (2025). Agro-industrial waste upcycling into activated carbons: A sustainable approach for dye removal and wastewater treatment. *Sustainability, 17*(5), 2036.

Bekchanov, D., Mukhamediev, M., Yarmanov, S., Lieberzeit, P., & Mujahid, A. (2024). Functionalizing natural polymers to develop green adsorbents for wastewater treatment applications. *Carbohydrate Polymers, 323*, 121397.

Chahal, M., Kumari, S., Bhattacharya, A., & Garg, M. C. (2024). Evaluating sustainable agricultural waste biomass for methylene blue adsorption in wastewater treatment: A state-of-the-art review. *Bioresource Technology Reports, 28*, 101983.

Duwiejuah, A. B., Payne, J., & Yahaya, D. (2024). Adsorption of toxic metals from landfill leachate using orange and banana peel powders as adsorbents. *Sustainable Environment, 10*(1), 2405295.

Iyer, S., Deshmukh, S. M., & Tapre, R. W. (2024). Review on removal of heavy metals from industrial effluents by adsorption. *Reviews in Inorganic Chemistry*. https://doi.org/10.1515/revic-2024-0079

Jahin, H. S., Khedr, A. I., & Ghannam, H. E. (2024). Banana peels as a green bioadsorbent for removing metals ions from wastewater. *Discover Water, 4*(1), 36.

Kumar, S., Manokar, S. N., Thirunavookarasu, N., Nivethitha, V., Nidhusri, T. N., Niranjana, T., & Rawson, A. (2022a). Characterization of power ultrasound modified kappaphycus alvarezii biosorbent and its modeling by artificial neural networks. *Water, Air, & Soil Pollution, 233*(7), 234.

Kumar, S., Thirunavookarasu, N., & Rawson, A. (2022b). Development of power ultrasound-modified industrial tomato waste as an efficient biosorbent: Characterization, application on synthetic dyes, and optimization using artificial neural networks. *Journal of Food Processing and Preservation, 46*(11), e17041.

Kumar, U., Gautam, V., & Prasad, K. (2024). Low-cost chemically treated sugarcane bagasse for removal of cationic, anionic and neutral dyes from aqueous solution. *Indian Chemical Engineer, 67*(2), 143–159.

Mubarak, A. A., Ilyas, R. A., Nordin, A. H., Ngadi, N., & Alkbir, M. F. M. (2024). Recent developments in sugarcane bagasse fibre-based adsorbent and their potential industrial applications: A review. *International Journal of Biological Macromolecules, 277*, 134165.

Nyairo, W., Njewa, J. B., & Shikuku, V. O. (2025). Adsorption of heavy metals onto food wastes: A review. *Frontiers in Environmental Chemistry, 6*, 1526366.

Okoro, H. K., Emenike, E. C., Iwuozor, K. O., Egbemhenghe, A., Bello-Hassan, M. T., Adu, A. O., & Adeniyi, A. G. (2025). Industrial waste biochar for heavy metal and dye remediation in wastewater: An overview. *Water Practice Technology, 20*(3), 595–616.

Sharma, P., Sharma, S., Sharma, S. K., Jain, A., & Shrivastava, K. (2024). Review on recent advancement of adsorption potential of sugarcane bagasse biochar in wastewater treatment. *Chemical Engineering Research and Design*. https://doi.org/10.1016/j.cherd.2024.04.055

Tariq, A., Yahaya, N., & Sajid, M. (2024). Low cost adsorbents derived from vegetables and fruits: Synthesis, properties, and applications in removal of heavy metals from water. *Desalination and Water Treatment, 320*, 100626.

Tolkou, A. K., Maroulas, K. N., Theologis, D., Katsoyiannis, I. A., & Kyzas, G. Z. (2023). Natural peels or peels-based activated carbons for the removal of pharmaceuticals, dyes, heavy metals, and anions. *Membranes, 10*, 22.

Tolkou, A. K., Tsoutsa, E. K., Kyzas, G. Z., & Katsoyiannis, I. A. (2024). Sustainable use of low-cost adsorbents prepared from waste fruit peels for the removal of selected reactive and basic dyes found in wastewaters. *Environmental Science and Pollution Research, 31*(10), 14662–14689.

Verma, S., Kumar, N., Kim, K. H., & Park, J. (2024). Photocatalytic efficacy of air purifiers equipped with self-cleaning titanium dioxide xerogel coatings against gaseous formaldehyde: A study using DRIFTS and DFT analysis. *Chemical Engineering Journal, 486*, 150269.

2 Characteristics of Biomass and Agro-Industrial-Based Materials

Shwetha Papani and M. S. Shwetha

2.1 INTRODUCTION

Renewable energy sources are crucial in present global initiatives to mitigate greenhouse gas emissions and to some extent, substitute conventional non-renewable energy sources. Climate change resulting from the emission of greenhouse gases (GHGs) represents just one facet of the numerous issues stemming from an unsustainable linear economy. These issues encompass escalating prices of raw materials, growing demand for products, depletion of resources, irreversible environmental harm, and the accumulation of waste. Consequently, a novel perspective on regenerative economics, centered on achieving a harmonious equilibrium among the economy, environment, and society, advocates for a circular production/consumption system. This system strives to maximize resource utilization while minimizing environmental impact [1].

Embracing a circular approach involves substituting fossil fuels and petroleum-based products, such as plastics, with sustainable energy systems (e.g., solar, wind, and biofuels) and renewable feedstock, predominantly bio-based products. It also entails the development of products that are easily recyclable or biodegradable, the optimization of waste management systems, and the design of products derived from residues [2]. Among these sources, biomass stands out as the most prevalent. The utilization of renewable energy is gaining significance in addressing global warming and ensuring a sustainable fuel supply. Over the last decade, there has been a resurgence of worldwide interest in biomass as a compelling substitute for fossil fuels. This interest is particularly pronounced in developing nations where biomass is abundant, and with effective management, it serves as a renewable resource, offering a cost-effective indigenous fuel supply [3].

Developing intelligent energy systems centered on biomass as the primary energy source provides opportunities to improve benefits in terms of the economy, environment, and social aspects. Exploiting the potential of biomass for heating within municipal and residential sectors should encompass various biomass fuels. This

DOI: 10.1201/9781003767251-2

includes biomass fuel that has been compressed, biomass fuels that have a significant bulk density but a low heating value, and biomass waste that is produced during the manufacturing of food with a high calorific value. The strategy entails utilizing fuel power units that are both efficient and low in emissions, specifically designed for small-scale purposes like individual agricultural properties and community heating systems. Utilizing biomass as a fuel source has minimal negative effects on the environment. When biomass is burned, it generates significantly lower amounts of NO and SO_2 compared to the combustion of non-renewable energy sources. Compared to petrol and diesel, liquid biomass fuels like 'ethanol and methanol' emit fewer 'carbon monoxide, hydrocarbons and possibly hazardous substances' into the atmosphere. Using fuel derived from biomass does not increase the amounts of carbon dioxide that cause global warming, in contrast to burning fossil fuels [4].

Bioenergy emerges as a sustainable and eco-friendly substitute for fossil fuels within the realm of renewable energy sources, leading to widespread international research. The domain of biomass conversion processes has swiftly grown due to the imperative to tackle the increasing energy shortfall and achieve a 70–90% reduction in CO_2 emissions. Biomass is predominantly transformed into biofuels through two key methods: 'thermo chemical and biochemical processes'. 'Thermo chemical conversion entails heating biomass material, either in the presence (gasification) or absence of a reactive environment (pyrolysis). This results in the production of gaseous (volatiles), solid (char) and liquid (tar) products'. On the other hand,

> biochemical conversion entails processing biomass through biological or chemical ways to produce bio-fuels. While biochemical conversion provides unique and discerning choices in processing which are not found in any other methods, it is inherently possessing greater complexity and cost compared to the thermo chemical route [5].

2.2 AGRO-INDUSTRIAL-BASED MATERIALS

In the 21st century, there is a significant difficulty in efficiently using biomass derived from agro-industrial wastes to produce 'food, drugs, biomaterials, biologically active chemicals and sustainable energy'. The ago-industrial produce significant amounts of waste, presenting an opportunity to repurpose these by-products as raw materials to create valuable products. This opens up prospects for sustainable production practices. Moreover, around 40–50% of the total volume of discarded food comprises fruits, vegetables, roots, and tubers. Given this situation, making efficient use of these waste resources has become imperative for producing goods in a way that is sustainable. Various agro-industrial wastes, including peel, seed, pit, pulp, press cake, and leaves, are under extensive investigation. These residues serve as sources for a broad spectrum of secondary plant metabolites, particularly phenolic compounds recognized as a notable category of bioactive compounds exhibiting antioxidant properties present in the tissues of fruits. The primary components of most plant materials are 'cellulose, hemicellulose and lignin'. 'Lignocellulosic materials, which include agricultural residues, forestry by-products, grasses, and woody materials, present considerable potential for the generation of bio fuels'. Additionally, agro-industrial wastes open avenues for exploration in other vital research domains, including the development of novel fuels and the production of energy. Furthermore, polymeric cellulose derived from these wastes can contribute to the creation of innovative biomaterials such as 'bio-nano composites, bio films, bioaerogels, hydrogels and specific biomaterials applicable in tissue engineering' [5, 6].

A growing demand is being shown in the benefits of bio-sourced biomass from lignin in recent years, due to global economic and environmental pollution concerns. Agro-industrial biomass, which consists of lignocellulosic waste, is a cost-effective, renewable, and plentiful resource that offers a unique opportunity for large-scale bioenergy production. Biotechnology, with its rapidly advancing tools, can help reduce conversion costs and increase the yield of desired products, thereby expanding the utilization of natural bioresources. Green biotechnology, in particular, shows great potential for converting solid farm wastes, particularly lignocellulosic materials, into biofuels in liquid form. Significant progress has already been made, positioning cellulosic ethanol as a competitive alternative to corn ethanol. Consequently, the vast quantities of lignocellulosic biomass have the potential to be transformed into various high-value products, including biofuels, value-added fine chemicals, and cost-effective energy sources for microbial fermentation and enzyme production [7].

2.3 CHARACTERISTICS OF BIOMASS

Biomass refers to the residual matter from animal waste or substances derived from plants, such as wood, charcoal, and agricultural residue, including algae. Plants absorb solar energy through photosynthesis, facilitating their growth and storing this energy in the form of biomass. Biomass may be used as a fuel in combustion processes to capture this energy. There is a chance that this will reduce dependency on conventional fossil fuels. Biomass is suited for use in combustion systems to produce electricity and heat energy because it may be converted into 'liquid, gaseous and solid states' by a variety of 'chemical, physical, and biological processes'. Currently, biomass accounts for about 14% of the world's total energy consumption, making it the fourth largest energy source globally. Biomass is a primary energy source for many developing countries, where it constitutes 20–33% of energy use, while in industrialized nations, it makes up between 9 and 14% of all energy supply. Biomass use can account for between 50 and 90% of total energy demand in certain developing nations. The majority of biomass is used for cooking and heating rather than for commercial reasons among the impoverished populations in these nations. Modern biomass energy production is used in industrialized nations to provide heat for heavy industries, transportation fuels, and electricity. The demand for biomass energy has significantly increased in developed regions like the European Union over the past few decades [8].

Biomass is the collective word for materials obtained by cultivating vegetation, including 'algae, crops, trees and plants', as well as waste from animal dung. Through the process of photosynthesis, plants take in solar energy and use it to grow and store it as biomass. Biomass energy can be recovered by burning these materials and using them as fuel [9, 10]. Biomass energy is recognized as a significant component of today's energy sources, offering a means to reduce dependence on petroleum products, natural gas, and coal. The primary objective of transforming biomass materials is to convert carbonaceous solids, which are initially difficult to handle due to their bulkiness and low energy density, into liquid fuels with properties that facilitate easy and cost-effective storage and transport through pumping systems.

Utilizing biomass products provides substantial environmental benefits. During growth, biomass materials absorb carbon dioxide and release it during combustion. Consequently, atmospheric carbon dioxide is recycled without creating a net warming impact. 'Agricultural waste (5%), municipal solid waste (24%) and landfill gases (5%), after which comes from wood and its wastes (64%), are the main sources of biomass energy'. The method of creating fuel from biomass is both economically viable and, when done with annual crops, may even have a good impact on the environment. One of the most traditional and adaptable ways to turn waste into energy is still combustion [8].

2.4 PYROLYSIS CHARACTERISTICS OF BIOMASS AND ITS COMPONENTS

India's economy is heavily reliant on agriculture, so the prospect of transforming abundant biomass resources into various forms of fuel holds significant allure. In various thermochemical conversion pathways, pyrolysis characteristics play a crucial role. Extensive research conducted in the last few decades has concentrated on comprehending the impact of operational factors such as 'temperature, heating rate, pressure and residence time' on pyrolysis (Table 2.1). Despite various types of biomass sharing common major components, their compositions can vary. The effects of 'biomass composition, ash content and composition on pyrolysis' have been explored. Some efforts have been made to establish connections between the pyrolysis traits of biomass and its constituents, including 'cellulose, hemicellulose, lignin and extractives' [11].

TABLE 2.1
Pyrolysis Characteristics of Biomass Components

Components	Yield		Temperature at Maximum Rate (K)	Initial Temperature at Decomp (K)	Temperature at Involution Point (K)
	Volatiles	Char			
Cellulose	97.5	2.5	682	573	713
Wood cellulose	86	14	667	573	697
Alkali lignin	59.4	40.6	695	413	773
Acid lignin	52.9	47.1	693	473	700
Hemicellulose	68	32	573	448	550
Extractives	73	26.9	668	393	575

TABLE 2.2
Pyrolysis Properties of Biomass

By-Products	Yield		Temperature at Maximum Rate (K)	Initial Temperature at Decomp (K)	Temperature at Involution Point (K)
	Volatiles	Char			
Bagasse	79.7	20.5	677	483	688
Cashewnut shell	81.1	18.9	638	473	673
Coconut shell	70.7	29.3	615	518	678
Corn cob	73.5	26.5	603	533	653
Cotton gin waste	80.6	19.4	679	523	688
Millet husk	70	29.9	653	523	663

2.5 CHEMICAL PROPERTIES OF BIOMASS AND ITS COMPONENTS

The use of compacted biomass fuel alone should not be the only way that biomass is utilized for heating in residential and municipal contexts. It should also contain biomass fuels with low bulk density and heating value, as well as waste biomass produced during the manufacturing of food with a higher calorific value. Small-scale, low-emission fuel power units should be used for this purpose (Table 2.2). They should be efficient. Ash is a by-product resulting from the combustion of biomass. The waste generated during biomass burning in households is often haphazardly collected and sent to landfills, posing an unnecessary environmental burden. The components of biomass fuels that cause ash vary significantly between different fuels and even within the same fuel type. Recognizing the ash characteristics can enable its application in agriculture for fertilization. Biomass ash comprises varying in proportion of 'silicon, aluminum, iron, calcium, magnesium, sodium, potassium, titanium and manganese', with impurities such as 'heavy metals'. The most eco-friendly and sustainable disposal method involves returning the ashes from biomass combustion to the soil. This approach allows a substantial portion of the macro- and micro-nutrients absorbed by plant parts to be reintroduced to the soil, completing the mineral circulation cycle. They may also include detrimental elements like heavy metals (nor, ender). The composition of ash is typically categorized based on its primary constituents in the mineral composition (Table 2.3). These components are further divided into three groups based on their levels of concentration, 'macro elements, including P, K, Ca and S; microelements, such as Mn, Fe, Cu and Zn; and toxic elements, which encompass Cr, Ni, As and Pb' [3, 12, 13].

2.6 PHYSICOCHEMICAL CHARACTERIZATION OF BIOMASS AND ITS COMPONENTS

Among the global challenges, the indiscriminate burning of fossil fuels and the resulting impact of CO_2 emissions on climate change are among the most critical

TABLE 2.3
Chemical Characteristics of Biomass and Biomass Components

By-Products	Ash	P	K	Ca	S	Cu	Fe	Mn	Zn	Pb
Pine wood	0.046	459.43	1419.16	3407.62	89.46	1.73	32.65	244.24	1.12	0.46
Oak wood	0.008	52.85	5433.49	3508.10	317.16	2.31	183.01	376.43	6.11	3.86
Wood residue chips	0.065	198.95	976.94	3949.64	204.58	1.00	33.65	172.42	1.00	0.53
Virginia mallow	0.090	349.16	1766.96	4580.64	137.79	1.47	92.70	16.77	3.00	0.54
Multiflorous rose	0.323	900.99	3791.34	5908.95	602.48	0.88	24.09	2.00	8.72	0.03
Wheat straw	0.212	157.82	849.32	611.31	80.55	0.32	45.83	4.04	2.08	0.07
Oat straw	0.673	130.31	939.36	1426.03	84.69	0.56	29.60	7.94	1.53	0.14
Barley straw	0.132	272.29	742.94	106.57	17.24	0.73	30.27	4.58	1.00	0.12
Beech bark	0.126	859.23	407.75	14,448.62	46.70	2.65	125.13	13.43	1.96	0.08
Hornbeam bark	0.121	206.61	379.75	4727.50	935.15	0.89	36.10	42.92	5.03	0.30
Cherry pits	0.050	118.07	3729.06	2385.99	63.51	11.15	145.57	20.26	11.37	0.07
Sunflower husk	0.14	428.64	1806.89	3094.12	117.63	0.75	31.90	8.14	1.73	0.10
Apple pomace	0.266	176.21	3637.03	400.83	64.50	24.56	155.18	8.39	29.02	1.23

issues of the 21st century. Among renewable energy sources, bioenergy stands out as a sustainable and environmentally friendly alternative to fossil fuels, prompting significant research efforts worldwide. The field of biomass conversion processes has rapidly expanded in science and technology, aiming to address the increasing energy demand and reduce 'CO_2 emissions by 70–90%'. Biofuels are produced from biomass through two primary pathways: biochemistry and thermochemical reactions. Thermochemical conversion involves heating biomass material in a reactive atmosphere to produce 'gaseous (volatiles), solid (char) and liquid (tar) products'. This process is known as 'gasification or pyrolysis'. Conversely, biochemical conversion processes biomass and produces biofuels by use of chemical and/or biological pathways. While biological conversion presents distinct options for selective processing that are not found in other methods, it is inherently more costly and intricate than thermochemical conversion. To effectively utilize lignocellulosic biomass feedstock in various conversion processes, it is essential to conduct fundamental studies on its physicochemical properties. This understanding aids in comprehending the overall process, selecting suitable conversion methods, and designing process parameters for biofuel production. Chemically, lignocellulosic biomass is highly heterogeneous, with its composition varying significantly depending on plant type, species, geographical location, age, growth conditions, and weather [14].

2.7 CHARACTERISTICS OF AGRO-INDUSTRIAL-BASED MATERIALS

The food and agriculture industries produce a significant quantity of trash, which can be turned into high-value products by using it as a raw material. This opens up a number of options for sustainable manufacturing. The bulk of agro-industrial wastes are dumped in landfills or are merely thrown in inappropriate locations, which adds to the rise in biomass from different processing sectors. The creation of new fuels and energy generation are two other significant research areas that are facilitated by agro-industrial wastes. The three main components of all plant materials are cellulose, hemicellulose, and lignin. Biofuel production can be greatly enhanced by lignocellulosic resources, such as woody materials, grasses, forestry leftovers, and agricultural wastes. Generally, '10–25% lignin, 20–30% hemicellulose and 40–50% cellulose' make up the majority of agricultural lignocellulosic biomass. The main structural element of plant cell walls that gives them their mechanical strength is cellulose. Conversely, repeating polymers of pentoses and hexoses are frequently seen in hemicellulose macromolecules. Coniferyl alcohol, sinapyl alcohol, and p-coumaryl alcohol are three aromatic alcohols found in lignin that are generated by a biosynthetic process. These alcohols act as a barrier between cellulose and hemicelluloses [1, 5, 14].

2.8 BIOLOGICAL SIGNIFICANCE OF LIGNOCELLULOSIC BIOMASS

Based on a biotechnology perspective, lignocellulosic sources of biomass come in a variety that may be transformed into important bio-products like bioethanol and other biofuels. There have been notable developments in sustainable technological advancement pertaining to lignocellulosic biomass in recent years. In order to slow down global warming, there is a growing need to investigate alternate, affordable, and environmentally sustainable biofuel sources due to the growing expense of petroleum-based fuels and their negative effects on the environment, notably their contribution to GHGs. Environmental contamination, global warming, and the sustainability of extraction of oil in the future have raised public and business interest in natural renewable resources as renewable energy alternatives. A viable method for producing bioethanol economically is to use lignocellulosic or agro-industrial biomass, which contains carbohydrates that need to be converted into simple sugars like glucose before undergoing fermentation into ethanol.

For these reasons, nations all over the world are making investments in other forms of energy, with bioethanol serving as a key focus. Generally, there are many steps involved in the conversion of biomass made from lignin into more valuable goods like biofuels or fine chemicals, including pretreatment (mechanical, chemical, or biological), enzymatic hydrolysis, and fermentation [7].

2.9 PHYSICOCHEMICAL AND STRUCTURAL CHARACTERISTICS OF CELLULOSE

Cellulose, the most abundant renewable organic biopolymer, can be sourced from agro-industrial biomass, marine animals, and microorganisms. Lignocellulosic biomass, which is plant-based and environmentally friendly, has potential applications in the sustainable production of biofuels, biochemicals, bioethanol, and other high-value-added products. Typically, lignocellulosic biomass comprises 30–50% cellulose, 15–35% hemicellulose, and 10–20% lignin. The high cellulose content makes biomass an excellent source of nano-sized cellulose, which is highly valuable as a reinforcing material. The highly stable polymer known as cellulose is made up of glucose chains that can have approximately 12,000 residues connected to them. This is mostly made up of '(1,4)-D-glucopyranose units, which have an average molecular weight of about 100,000 and are joined by b-1,4 links'. Forty to fifty percent of the cellulose molecules in plant biomass are held together by intermolecular hydrogen bonds in their native state. However, these molecules also have a strong tendency to form intra- and intermolecular hydrogen bonds, which makes the cellulose more rigid and highly resistant to most organic solvents. In their natural state, cellulose molecules are found in bundles that combine to create microfibrils, which are divided into crystalline and amorphous regions [1, 5, 7, 14].

2.10 PHYSICOCHEMICAL AND STRUCTURAL CHARACTERISTICS OF HEMICELLULOSE

The second most common heterogeneous polymer is hemicellulose, which is primarily composed of 'glucuronoxylan, glucomannan and traces of other polysaccharides'. Mannan is a part of the hemicellulose found in both hardwood and softwood, while arabinan, galactan, and xylan are found in straws and grasses. They are categorized using xylans, mannans, and glucans as the structural backbone of sugars, with xylans and mannans being the most prevalent. The hemicellulose group includes galactans, arabinans, and arabinogalactans; 'however, they do not have the same equatorial b-1, 4 linked backbone structure. Glucuronoxylan, also known as O-acetyl-4-O-methyl-glucurono-b-Dexylan, is the main constituent of hardwoods'. The polymer's backbone, xylopyranose, is joined by b-1,4 links. Hemicellulosic biomass has an average molecular weight of less than 30,000 and 25–35% of hemicelluloses. Cellulose and hemicellulose are tightly bound to each cellulose microfibril through non-covalent attractions. Hemicelluloses were initially thought to be in the middle of the cellulose synthesizing process [5, 7]. Hemicelluloses represent the primary sugar component that comes in second after cellulose in lignocellulosic substances. These heteropolysaccharides consist of hexosan and pentosan monomer units. Typically comprising 20–30% of wood, softwood hemicelluloses include 'both pentosans and hexosans, whereas hardwood hemicelluloses are primarily pentosans. Most hemicelluloses have a low degree of polymerization, generally around 100–200, and they are branched structures that lack crystalline regions. Their complex composition and linkage to lignin pose significant challenges'. Hemicelluloses are essential because they act as a link between 'lignin and cellulose'. They form hydrogen bonds with cellulose and covalent bonds which might be ester, ether, or glycosidic in nature with lignin. Approximately, 80–90% of the lignin and 50% of the hemicelluloses are removed during standard chemical pulping procedures. During the recovery process, the dissolved lignin and the broken down hemicelluloses, mostly aldonic acids, in the wasted pulping liquid are usually burned together. But as a stand-alone fuel source, wood hemicelluloses are less efficient since their thermal energy value about '13.6 MJ/kg is only around half that of lignin'. As a result, direct combustion presently utilizes hemicelluloses insufficiently. In the context of pulp mill biorefineries, a more economically viable approach involves extracting hemicelluloses before pulping and converting them into 'higher-value products such as ethanol, itaconic acid or other chemicals'. This strategy aims to maximize the utilization of hemicelluloses and enhance the overall efficiency and sustainability of the pulp and paper industry [10].

2.11 PHYSICAL AND STRUCTURAL CHARACTERISTICS OF LIGNIN

Lignin typically makes up between 10 and 25% of the biomass by weight, making it the smallest and most complicated portion. Its polymer is heterogeneous and long-chain, primarily made up of phenylpropane units, which are typically joined by ether bonds. By enclosing the cellulose and hemicellulose molecules with the polymers, lignin functions as a glue. Since it may be found in any plant biomass, it is regarded as a by-product or residue in the process of producing bioethanol (Table 2.4). Lignin, which binds the components of cell walls together, is made up of a 'complex and massive polymer of phenylpropane, methoxy groups and non-carbohydrate polyphenolic material' [15]. These phenylpropanes, identified by 0, I, and II methoxyl groups attached to their rings, form unique structures I, II, and III. These structures vary

TABLE 2.4
Percent Composition of Agriculture Wastage (Physical and Structural Characteristics)

S. No	Agriculture By-products	Cellulose (%)	Hemicellulose (%)	Lignin (%)
1	Sugar cane bagasse	42	25	20
2	Sweet sorghum	45	27	21
3	Hardwood	40	24–40	18–25
4	Corn stover	38	26	19
5	Rice straw	32	25	18
6	Nut shells	25–30	25–30	30–40

based on the plant source: structure I is found in grasses, structure II in coniferous wood, and structure III in deciduous wood [5, 7].

The combination of hemicellulose and lignin forms a protective barrier around cellulose, which needs to be modified or removed to facilitate efficient hydrolysis of cellulose. The crystalline structure of cellulose makes it highly insoluble and resistant to enzymatic attack. Therefore, compared to processing sugar or starch crops, more advanced pre-treatment technologies are necessary for economically hydrolyzing hemicellulose and cellulose. Once cellulose and hemicellulose are converted into sugars (saccharified), the subsequent ethanol production process is similar to that used for grain-based ethanol.

2.12 PROPERTIES OF BIO-WASTE DERIVED FROM AGRO-INDUSTRIAL AND ITS BY-PRODUCTS

When a plastic material possesses both bio-based and biodegradable qualities, it is referred to as bioplastic. The phrase 'bio based' indicates that the majority of its constituent parts come from biomass, whereas 'biodegradation' refers to the chemical process by which a substance, in the presence of normally occurring environmental circumstances, transforms into water, carbon dioxide, and compost through the action of naturally occurring microbes. Current research trends are exploring the potential of natural materials, like fibers, in construction products, highlighting their numerous advantages. These include minimal environmental impact, reduced energy requirements, affordability, lightweight properties, widespread availability, biodegradability, and excellent insulation and mechanical characteristics. Numerous studies have investigated the utilization of bio-wastes in construction materials, assessing various applications and analyzing their diverse properties. For instance, research has compiled data on natural and recycled materials utilized specifically for thermal and acoustic insulation purposes.

- Chemical properties: organic matter must make up at least 50% of the material's final composition.

- Biodegradation: after six months, under stimulated composting conditions, it should decompose to at least 90% of its original weight or volume.
- Ecotoxicity: plant development should not be endangered by non-biodegradable residues that have been biodegraded for six months.
- Disintegration: in controlled composting circumstances, the microscopic pieces of the components should be undetectable (<2 mm) after at least two months.

Then, bioplastics (both biodegradable and non-biodegradable) can be divided into four primary categories: those that are made from microorganisms, directly derived from biomass, bio-based monomers, and petrochemicals [2].

2.13 FOREST MANAGEMENT AND FLORICULTURE

Previous studies examined the application of bio-wastes from floriculture and forest management. The possibility of producing thermal insulation materials from Hortensia (*Hydrangea macrophylla*) stems was investigated. The scientists looked into the chemical and physical characteristics of the plant, as well as a composite material that included commercial polyurethane. The outcomes showed that when compared to commercial expanded polystyrene, it performs better and has greater insulation qualities [4, 16].

According to their analysis, particleboards composed of cattail fiber (Typha), a plant that grows quickly and negatively affects other crops, perform well because of their poor thermal conductivity (between 0.044 and 0.061 W/(m.K)). Various bio-based aggregates made from Brassica, such as rape straw. They examined its chemical and physical characteristics and found that it had high water absorption, low thermal conductivity, good hygrothermal qualities, and a moisture buffer value. The results of water absorption showed a high porosity, which is crucial for effective thermal insulation. Indeed, there is an inverse correlation between open porosity and thermal conductivity. The moisture buffer value indicated a material's capacity to regulate interior conditions passively [17].

2.13.1 Agricultural Practices

In order to assure crop growth or harvesting, agriculture produces a lot of bio-wastes, which are frequently discarded yet can be a valuable resource for building. Previous studies have shown that these bio-wastes can be used for a variety of purposes, taking into account both composites and raw materials. The leftovers from agricultural techniques are used to make a variety of building products. Researchers looked at several applications of the raw materials, including the behavior of a straw bale wall in terms of heat and sound, conducting both in situ and lab analyses. Results from the experiment showed good thermal performances: the samples' thermal conductivity values, which fell between 0.050 and 0.060 W/(m.K) at a density of 80 kg/m^3, were below the maximum advised for thermal insulators, which is 0.065 W/m.K [18].

2.13.2 Agro-Industrial Processes

A major problem of the 21st century is the long-term sustainability of biomass from agricultural and industrial waste to produce food, medicines, biomaterials, chemically active substances, and renewable energy sources. Large amounts of trash are produced by the food and agriculture sectors, which may be used as raw materials to make valuable products. This strategy encourages a variety of chances for environmentally friendly production methods. Among natural materials, bio-wastes offer a practical choice for usage as construction materials. As mentioned before, they are mass-produced, readily available and may be used to make a variety of composites. A substantial amount of agricultural and industrial waste from different industries winds up in landfills or is disposed of incorrectly, which contributes considerably to the buildup of biomass. Since this increased biomass comprises highly energetic organic components like glucose and lipid polymers, it poses problems for both microbiological activity and fuel potential. In arid areas, biomass buildup can intensify forest fires, resulting in significant losses of ecosystems and forests. Furthermore, these wastes serve as a food supply for microbes, which are beneficial for the breakdown process but can cause problems on a big scale when they produce harmful by-products, GHGs, and infections.

In developing countries, the abundant agro-industrial waste often lacks effective substitutes for recovering and recycling, resulting in practices like incineration that increase greenhouse gas emissions, waste energy potential, exacerbate health crises, and impact quality of life. Although agro-industrial waste contains valuable organic matter, few technologies currently ensure 'secure recovery, purification and concentration of these chemical compounds'. Moreover, there is a scarcity of technologies in development that provide operational data applicable to scaling up these processes. Consequently, there is a growing focus on exploring innovative technologies to address these challenges and promote the sustainable reuse of waste. In order to promote possibilities for research and technology development in the pursuit of environmentally friendly waste disposal techniques, the goal is to integrate these technologies within industrial settings.

Based on studies of construction products, bio-wastes can be classified into several types: those from forest management, floriculture, and the use of flowers and shrubs; those from agricultural and manufacturing practices involving agro-products; and those from the food industry. The global production and consumption of these bio-wastes are significant and growing, making them an excellent source for producing natural composites. The intake and use of farm produce in manufacturing processes are another source of bio-waste; several studies have examined their properties to encourage their re-use. Many researchers examined the performance of a composite of hemihydrate gypsum and tea waste powder to produce plastering paste; the addition of tea waste resulted in longer setting times and higher density; initially, the sample without any tea waste had a density of 2.14 g/cm^3, but after adding 2.5% in mass, the sample with 2.5% in mass raised from '15 min and 22 min to 840 min and 3500 min', respectively [4, 19].

2.13.3 Pharmacological Application

A lot of study has been done using new technology, especially in the pharmaceutical industry, to extract useful components from agricultural and industrial waste. By synthesizing pharmaceuticals using non-conventional materials and methods, this strategy seeks to bring value to large-scale biomass production, particularly that of the food sector, while providing sustainable alternatives to established processes. Rather than being disposed of simply, agro-industrial wastes have been investigated in recent years for sophisticated and eco-friendly uses. For instance, 'the by-product of high-value agriculture, cocoa bean hulls, has a chemical makeup rich in phenolic chemicals that may find use in nutraceutical goods. Similarly, in individuals receiving treatment, grape seed oil has demonstrated encouraging results in raising HDL and lowering LDL cholesterol levels. Studies involving Wistar rats treated with grape seed oil for ten weeks demonstrated reduced cholesterol levels in the liver, attributed to decreased food intake' [20].

2.13.4 Application as Ingredients for Food Products

The agri-food industries produce large volumes of trash every day. These wastes come from several parts of the production chain, including 'harvesting, shipping, storage, industrial processing, commercialization and ultimate consumption'. Rich supplies of vital nutrients, including vitamins, minerals, fiber, carbohydrates, and protein, may be found in these residues, which include peel, seed, and other fruit pieces, frequently in higher quantities than in other fruit sections. Agro-industrial residues are excellent as 'high-value natural food components, additives, and supplements since they also include lipids, natural colors, enzymes, pigments, antioxidants, and antimicrobials'. The creation of novel food items is facilitated by the reuse of these wastes in industrial production, providing workable alternatives that are advantageous for the surroundings and for economic growth [21].

For example, studies have explored the potential of dry powders derived from fruit residues such as 'pomegranate, grapes, lemon, apple and papaya', as sources of natural pigments, yielding promising results. These pigments, derived from various fruit residues, contribute beneficial health effects. Commonly used as culinary coloring, carotenoids from tomato and pomegranate peels have important health advantages as precursors to vitamin A, strengthening immune systems and improving skin health. Similarly, carotenoids from melon residues contain valuable antioxidants like lutein and β-carotene, which promote health by combating oxidative stress.

Antioxidant compounds that are natural are becoming more and more in demand than synthetic alternatives because of their biocompatibility, low toxicity, cost-effectiveness, ease of production, and environmentally friendly disposal. Agro-industrial waste is also utilized in developing new food products with functional properties. For instance, ingredients for cookies have been developed using extracts from the peels of pomegranates enclosed with orange processing waste from manufacturing. 'This encapsulation process maintains the product's sensory quality, antioxidant activity, and oxidation stability while preserving phenolic components. Other high-value residues, such the residues from goji berries, have been effectively used to replace wheat flour in baked goods, adding proteins, free phenolic compounds, and soluble and insoluble dietary fibres'. These innovations demonstrate the potential of agro-industrial waste as functional ingredients with excellent nutritional and sensorial properties, contributing to sustainable food production practices [22].

2.13.5 Utilization of Lignocellulosic Wastage for the Production of Biomaterials

Organic, sustainable, and ecologically beneficial matrices, lignocellulosic raw materials are widely accessible at a cheap cost and have great potential for use in a wide range of industrial applications. When compared to synthetic inputs, these materials have a number of benefits, including the ability to be sourced from a variety of sources, such as forests and agro-industrial leftovers; lower density and the energy requirements; specific strength and modulus; easily modifiable reactive surfaces; and suitability for recycling and nanoparticle systems. By weight, lignocellulosic biomass usually consists of 40–50% cellulose, 20–30% hemicellulose, and 10–25% lignin. The main non-carbohydrate component of plant cell walls, lignin, gives them structural strength and hydrophobicity, which protects carbohydrates from microbial destruction. It is a complicated aromatic biopolymer molecule made up of 'hydroxyphenyl, guaiacyl and syringyl subunits'.

Hemicelluloses, another component, are polysaccharides that contribute to the cell wall composition of plants. They consist of various monomeric units besides cellulose and encompass a diverse class of linear and branched polysaccharides, including pentoses and hexoses. Cellulose, a major component, is a linear homopolymer with high molecular weight and is insoluble in water. It is composed of β-D-glycopyranosyl monomers linked by glycosidic bonds. The molecules of cellulose can align and bond through hydrogen bonds extensively, forming fibrous and polycrystalline structures. In summary, lignocellulosic biomass offers a robust foundation for sustainable material development and industrial applications, characterized by its composition of lignin, hemicellulose, and cellulose, each contributing unique properties that are essential for various manufacturing and environmental solutions [23].

Structured materials called bioaerogels are distinguished by their low density, large surface area, and micro- or nanoporous structure. Using maize straw, researchers have created spongy aerogels that efficiently absorb oil. These biodegradable compounds were synthesized utilizing the sol–gel technique in a water-in-oil emulsion system, using procedures including shearing and lyophilization. Due to their unique features, the final product is ideal for use in controlled drug delivery systems. Similarly, rice straw aerogels show promise for oil separation processes with features including exceptional amphiphilic absorption qualities, ultra-lightweight, and ultra-porosity.

A biofilm was created by combining glucosamine and small silver particles with lignocellulosic components from bagasse used in sugar production. This compound demonstrated improved swelling characteristics and flexibility, especially at high lignin concentrations. The resultant hydrogel showed antioxidant and antibacterial properties. Furthermore, research was conducted on a nano-biofilm that used nanocellulose biopolymer made from plant biomass, notably maize leaf wastes from agro-industrial processing that were not chemically treated. Numerous factors were examined, including 'mineral composition, form, stiffness, tensile strength, pH, cellulose content, water absorption and odour'. This novel substance has potential as a biodegradable substitute for plastics made from petroleum. It emphasizes how natural, renewable resources may be used to generate sustainable materials with qualities appropriate for uses where environmental compatibility and biodegradability are critical factors [24].

2.13.6 Agricultural and Industrial Waste as a Substitute Energy Source

Agricultural crops play a crucial role in global food production and have significant socio-economic impacts, influencing income distribution and contributing broadly to the global economy. Over the past decade, crop products such as cereal, legume, oilseed, sugar, and tuber have seen substantial growth, leading to increased agro-industrial waste production. It is estimated that between 25% and 30% of global energy biomass could potentially come from these agro-industrial residues, representing a significant resource for sustainable energy production. Research efforts have focused extensively on converting non-woody biomass into energy sources through processes like pyrolysis for oil production, which is then used for combustion to generate heat and energy. Wet biomass and other residues are utilized for producing hydrogen and biogas through fermentation processes, both solid-state and submerged, as well as anaerobic digestion.

Biomass, which is rich in sugars and polysaccharides such as 'starch, pectin, cellulose and chitin', is particularly suitable for bioethanol production using enzymatic hydrolysis, saccharification, and fermentation processes. These technologies offer promising alternatives considering their technological feasibility and economic viability. Various methods are employed in biomass energy production, each offering different forms of energy. Conventional combustion processes convert biomass into electricity by burning it in boilers to produce high-pressure steam, which drives turbines for electricity generation. Gasification technology is also utilized to convert agro-industrial waste into mechanical work and electricity. The development of these energy production processes underscores the need for substantial investment in research and technology. Further studies are essential to explore combined processes, extraction and purification methods, and the potential use of supercritical fluid CO_2 as a catalyst for the generation of energy. These advancements are crucial for enhancing the efficiency and sustainability of biomass-based energy systems [25].

2.14 CONCLUSION

In conclusion, when examining the characteristics of biomass and materials derived from agro-industrial sources, it is evident that these resources offer diverse and sustainable alternatives for various applications. The renewable nature, abundance, and versatility of biomass make it a promising option for energy production, as well as the development of eco-friendly materials. Additionally, agro-industrial by-products showcase the potential to contribute to the creation of innovative and sustainable materials, reducing waste and environmental impact. Embracing these characteristics opens up avenues for creating a more sustainable and environmentally conscious future in the energy and materials industries.

Biomass and agro-industrial materials represent versatile resources with distinctive characteristics that make them pivotal in sustainable development and various industrial applications. These materials encompass a wide array of organic matter derived from agricultural and food processing activities, including residues like peels, seeds, husks, and other by-products. One of the key features of biomass and agro-industrial materials is their abundance and renewable nature, making them sustainable alternatives to finite fossil fuels. In terms of composition, biomass typically consists of cellulose, hemicellulose, lignin, and other organic compounds. Cellulose provides structural integrity, while hemicellulose and lignin contribute to the complex matrix that gives biomass its strength and resilience. These materials are rich in carbohydrates, proteins, fats, fibers, vitamins, minerals, and bioactive compounds, offering diverse opportunities for value-added products in the pharmaceuticals, food, cosmetics, and energy sectors.

Moreover, biomass and agro-industrial materials exhibit low environmental impact throughout their lifecycle. They require minimal energy for production and processing compared to synthetic counterparts and contribute to carbon sequestration when managed sustainably. Their biodegradability and potential for composting further underscore their eco-friendly attributes, aligning with global sustainability goals. Technological advancements have expanded the utilization of biomass and agro-industrial materials, enabling efficient extraction of valuable components such as biofuels, biochemicals, nutraceuticals, and biopolymers. Innovative approaches in biorefinery processes allow for the integrated and efficient conversion of these materials into multiple high-value products, contributing to economic growth and resource efficiency.

In conclusion, biomass and agro-industrial materials represent a valuable resource pool with diverse characteristics that support sustainable development and offer solutions to pressing environmental and economic challenges. Harnessing their potential through innovative technologies and responsible management practices is crucial for achieving a more sustainable and resilient future.

REFERENCES

1. Zou, R., Qian, M., Wang, C., Mateo, W., Wang, Y., Dai, L., Wang, L., Zhang, X., Kong, X., Ruan, R., & Lei, H. (2022). Biochar: From by-products of agro-industrial lignocellulosic waste to tailored carbon-based catalysts for biomass thermochemical conversions. *Chemical Engineering Journal*, *441*, 135972.
2. Ortega, F., Versino, F., López, O. V., & García, M. A. (2022). Biobased composites from agro-industrial wastes and by-products. *Emergent Materials*, *5*(3), 873–921.
3. Cuiping, L., Chuangzhi, W., & Haitao, H. (2004). Chemical elemental characteristics of biomass fuels in China. *Biomass and Bioenergy*, *27*(2), 119–130.
4. Cintura, E., Nunes, L., Esteves, B., & Faria, P. (2021). Agro-industrial wastes as building insulation materials: A review and challenges for Euro-Mediterranean countries. *Industrial Crops and Products*, *171*, 113833.

5. Teo, H. L., & Wahab, R. A. (2020). Towards an eco-friendly deconstruction of agro-industrial biomass and preparation of renewable cellulose nanomaterials: A review. *International Journal of Biological Macromolecules*, *161*, 1414–1430.
6. Freitas, L. C., Barbosa, J. R., da Costa, A. L. C., Bezerra, F. W. F., Pinto, R. H. H., & de Carvalho Junior, R. N. (2021). From waste to sustainable industry: How can agro-industrial wastes help in the development of new products? *Resources, Conservation and Recycling*, *169*, 105466.
7. Anwar, Z., Gulfraz, M., & Irshad, M. (2014). Agro-industrial lignocellulosic biomass a key to unlock the future bio-energy: A brief review. *Journal of Radiation Research and Applied Science*, *7*(2), 163–173.
8. Abuelnuor, A. A. A., Wahid, M. A., Hosseini, S. E., Saat, A., Saqr, K. M., Sait, H. H., & Osman, M. (2014). Characteristics of biomass in flameless combustion: A review. *Renewable and Sustainable Energy Reviews*, *33*, 363–370.
9. Demirbas A. (2007). Combustion of biomass. *Energy Sources Part A: Recovery, Utilization, and Environmental Effects*, *29*, 549–561.
10. Williams, A., Jones, J., Ma, L., & Pourkashanian, M. (2012). Pollutants from the combustion of solid biomass fuels. *Progress in Energy and Combustion Science*, *38*, 113e137.
11. Sardar, H., Waqas, M., Naz, S., Ejaz, S., Ali, S., & Ahmad, R. (2022). Evaluation of different growing media based on agro-industrial waste materials for the morphological, biochemical and physiological characteristics of stevia. *Cleaner Waste Systems*, *3*, 100038.
12. Nordin, A. (1994). Chemical elemental characteristics of biomass fuels. *Biomass and Bioenergy*, *6*(5), 339–347.
13. Zając, G., Szyszlak-Bargłowicz, J., Gołębiowski, W., & Szczepanik, M. (2018). Chemical characteristics of biomass ashes. *Energies*, *11*(11), 2885.
14. Saima, S., Nik, N., Rokiah, H., Khan, S. B., Widad, F., & Mohd, O. (2018). Physicochemical characterization of Malaysian crop and agro-industrial biomass residues as renewable energy resources. *Industrial Crops and Products*, *111*, 642–650.
15. Martins, R. S. F., Goncalves, F. G., de Alcantara Segundinho, P. G., Lelis, R. C. C., Paes, J. B., Lopez, Y. M., Chaves, I. L. S., & de Oliveira, R. G. E. (2021). Investigation of agro-industrial lignocellulosic wastes in fabrication of particleboard for construction use. *Journal of Building Engineering*, *43*, 102903.
16. Cárdenas-R, J. P. (2020). Thermal insulation biomaterial based on hydrangea macrophylla. In *Bio-based materials and biotechnologies for eco-efficient construction* (pp. 187–201). Woodhead Publishing.
17. Ratiarisoa, R. V., Magniont, C., Ginestet, S., Oms, C., & Escadeillas, G. (2016). Assessment of distilled lavender stalks as bioaggregate for building materials: Hygrothermal properties, mechanical performance and chemical interactions with mineral pozzolanic binder. *Construction and Building Materials*, *124*, 801–815.
18. D'Alessandro, F., Bianchi, F., Baldinelli, G., Rotili, A., & Schiavoni, S. (2017). Straw bale constructions: Laboratory, in field and numerical assessment of energy and environmental performance. *Journal of Building Engineering*, *11*, 56–68.
19. Huang, Y., Xu, C., Li, H., Jiang, Z., Gong, Z., Yang, X., & Chen, Q. (2019). Utilization of the black tea powder as multifunctional admixture for the hemihydrate gypsum. *Journal of Cleaner Production*, *210*, 231–237.
20. Ben Mohamed, H., Duba, K.S., Fiori, L., Abdelgawed, H., Tlili, I., Tounekti, T., & Zrig, A. (2016). Bioactive compounds and antioxidant activities of different grape (*Vitis vinifera* L.) seed oils extracted by supercritical CO2 and organic solvent. *LWT - Food Science and Technology*, *74*, 557–562.

21. Hosseini, S. E., Wahid, M. A. (2013). Feasibility study of biogas production and utilization as a source of renewable energy in Malaysia. *Renewable and Sustainable Energy Reviews*, *19*, 454–462.
22. Bellemare, M. F., Çakir, M., Peterson, H. H., Novak, L., & Rudi, J. (2017). On the measurement of food waste. *American Journal of Agricultural Economics*, *99*(5), 1148–1158.
23. Fortunati, E., Balestra, G. M. (2019). Lignocellulosic materials as novel carriers, also at nanoscale, of organic active principles for agri-food applications. In D. Verma, E. Fortunati, S. Jain, & X. Zhang (Eds.), *Biomass, biopolymer-based materials, and bioenergy* (pp. 161–178). Woodhead Publishing.
24. Aradmehr, A., & Javanbakht, V. (2020). A novel biofilm based on lignocellulosic compounds and chitosan modified with silver nanoparticles with multifunctional properties: Synthesis and characterization. *Colloids and Surfaces A: Physicochemical and Engineering Aspects*, *600*, 124952.
25. Prasad, S., Malav, M. K., Kumar, S., Singh, A., Pant, D., Radhakrishnan, S. (2018). Enhancement of bio-ethanol production potential of wheat straw by reducing furfural and 5-hydroxymethylfurfural (HMF). *Bioresource Technology Reports*, *4*, 50–56.

3 Synthetic Dyes, Heavy Metals, and Other Hazardous Materials for the Environment

G. H. Kavana, S Divya, Kamalika Sarkar, and Sumit Kumar

3.1 SYNTHETIC DYES: A BRIEF OVERVIEW

Synthetic dyes exert a considerable and extensive impact on industrial processes. Thanks to their extensive variety of vibrant, durable, and affordable color choices, these dyes have transformed several industries, such as food, textiles, and cosmetics. The accidental creation of Mauveine by William Henry Perkin in 1856 while attempting to formulate an anti-malarial drug is the origin of synthetic dye discovery. This accidental discovery revolutionized dyeing techniques and enabled the production of more affordable and efficient alternatives to natural dyes, which had previously been expensive and labor-intensive to produce (Hagan & Poulin, 2021). Synthetic dyes, known for their stability and adaptability across various applications as well as their bright color range, now dominate key industries. However, in spite of their clear industrial applications, the environmental and health impacts of synthetic dyes have raised significant concerns. Azo dyes represent a well-known group of synthetic coloring agents recognized for their harmfulness and persistence. These colors pose major health risks for humans and can severely harm ecosystems if not properly discarded. To reduce the negative impacts of dye pollution, recent studies emphasize the identification of environmentally friendly dye alternatives and enhancing the efficiency of wastewater treatment methods.

3.2 CLASSIFICATION OF SYNTHETIC DYES

Although synthetic dyes have given the industry brilliant and affordable color options, there are significant concerns about the effects they have on the environment and human health. The two primary categories of synthetic dyes, reactive and azo dyes, present significant difficulties for efficient waste management due to their toxicity, long environmental half-lives, and persistence.

 DOI: 10.1201/9781003767251-3

1. **Azo Dyes**

   Azo dyes, the most common and biggest family of synthetic dyes, are well known for having a detrimental impact on the environment. In addition to providing these dyes with their striking hue, nitrogen–nitrogen double bonds (-N=N-) also support their chemical stability and environmental endurance. Azo dyes are widely used in leather, textiles, and even food goods, but when they break down in low-oxygen areas like landfills or contaminated waterways, harmful byproducts like aromatic amines that cause cancer are produced. These amines can bioaccumulate in food chains and endanger human health in addition to directly endangering aquatic and terrestrial species (Ahmad et al., 2023).

   Ecosystem disturbance is one of the environmental dangers connected to azo dyes. Azo dyes produce pollution in streams, which is difficult to remove from the environment over time, in addition to lowering water quality. The extensive use of azo dyes remains a significant environmental concern, despite efforts to break them down more efficiently by photocatalysis, ozonation, and other cutting-edge techniques.

2. **Reactive Dyes**

   The high solubility of reactive dyes in water and subsequent challenging extraction from industrial effluent poses comparable challenges. These dyes ensure that colors in clothes are retained for a long time by creating covalent connections with fibers. Nevertheless, they are persistent in the environment due to the same chemical characteristics that make them strong on fabrics. Unreacted dyes in large amounts find their way into wastewater streams, seriously polluting neighboring rivers and lakes. Reactive dye-containing wastewater is especially difficult to treat as these substances are sometimes unable to be eliminated using standard wastewater treatment techniques (Cui et al., 2020). Serious environmental effects result from this, such as contaminated water that can harm ecosystems and communities of people who depend on these water supplies for agriculture or drinking.

3.2.1 Molecular Bonding Mechanisms in Synthetic Dyes

The molecular bonding processes responsible for the stability and industrial efficacy of synthetic dyes also give rise to their environmental persistence. A variety of single, double, and pi–pi bonds are used by other synthetic dyes to produce their vivid hues, but azo dyes, in particular, depend on a nitrogen–nitrogen double bond. These bonds give rise to dyes that are extremely resistant to natural degradation processes, particularly when they are a component of conjugated systems.

The issue with this chemical stability is that synthetic dyes are difficult for natural systems to decompose when they are discharged into the environment. Due to their prolonged persistence in soil, water, and even the air due to their resistance to breakdown, they cause long-term pollution and ecological damage (Ahmad et al., 2023). Instead of just disappearing, these dyes can build up in sediments, water bodies, and living things, which gradually makes them more poisonous.

3.2.2 Environmental and Health Impacts of Synthetic Dyes

Despite widespread acceptance, there are serious health and environmental risks related to the use of synthetic colors. Synthetic dyes have several detrimental effects that have a significant influence on both human health and the environment.

1. **Impact on Aquatic Eco-Systems**:
   Impacts on aquatic ecosystems are among the most serious environmental effects of pollution from synthetic dyes. Artificial dyes block light from penetrating water bodies, which prevents aquatic plants from photosynthesizing and lowers the water's oxygen content. Due to the absence of oxygen, this results in "dead zones," where aquatic life cannot thrive (Tripathi et al., 2023). The most susceptible species to this disturbance include fish, amphibians, and other aquatic life. Aquatic ecosystems are further contaminated by azo dyes because they break down and release carcinogenic aromatic amines. The long-term harm that these harmful substances may do to ecosystems makes them a serious threat to aquatic systems' biodiversity. These substances have the potential to reach the human food chain due to their bioaccumulation in aquatic creatures, which presents additional health risks.

2. **Soil Contamination and Groundwater Pollution**:
   Another major cause of soil pollution is synthetic colors. Industrial dyes are frequently disposed of in landfills or poured into drainage systems, where they affect natural ecosystems by seeping into the soil. (Ghais et al., 2020) highlight the particular worry regarding the dyes' effect on soil microorganisms, as they have the potential to modify the microbial equilibrium that is crucial to soil health and fertility. Particularly in locations close to industrial operations, this may result in lower agricultural productivity and crop yields. One more important problem is groundwater pollution. Aquifers and other subterranean water sources may get contaminated by artificial colors when they leak into the earth. These contaminants can remain for prolonged spans in groundwater and are challenging to eliminate using standard filtration methods, endangering human populations.

3. **Human Health Risks**:
   Synthetic dyes carry severe health hazards. When azo dyes break down, specifically, there is evidence that they release chemicals that might cause cancer. These substances—such as aromatic amines—have been classified by the International Agency for Research on Cancer as likely human carcinogens (Golka et al., 2004). Renal impairment, asthma, and several types of cancer are among the major health concerns that can result from prolonged exposure to these chemicals, whether through polluted food, water, or soil. Artificial food coloring, such as tartrazine and sunset yellow, is widely employed in the food business to improve the look of processed meals. Nevertheless, allergic responses, hyperactivity in youngsters, and other negative health consequences have been linked to these colors (Aguilar

et al., 2009). Similarly, in the cosmetics business, synthetic colors used in products such as hair dyes and beauty items have been connected to a higher risk of bladder cancer and other health difficulties (Wargala et al., 2021). Due to their extensive usage and difficulty in limiting exposure, these dyes continue to represent a serious health concern despite regulatory restrictions.

3.2.3 Waste Management and Sustainable Alternatives

Handling the trash produced by artificial coloring agents has emerged as a critical environmental issue. Chemical and physical filtration, two common techniques of treating wastewater, are sometimes not enough to eliminate synthetic colors from industrial effluent. According to Tripathi et al., (2023), the intricate molecular architectures of these dyes impede their breakdown, resulting in the discharge of hazardous compounds into the surroundings. Large-scale biological treatment techniques are useless because, although they are more ecologically friendly, they are also unable to dissolve the complex chemical linkages found in synthetic colors.

The limitations of existing waste management procedures have prompted research into more sophisticated approaches, such as ozonation and photocatalysis, which are examples of advanced oxidation processes (AOPs). Although these techniques are costly and have not yet gained widespread acceptance, they have the potential to convert synthetic dyes into less hazardous chemicals (Javaid & Qazi, 2019). Potential options also include membrane filtration and bioreactors, although these are not without cost and scalability issues. Industries are looking more and more into sustainable dye alternatives in response to the risks that synthetic dyes represent to the environment and human health. Although they are a more environmentally responsible choice, natural dyes made from plants, minerals, and microbes have drawbacks of their own, including increased prices and variable effectiveness.

Despite being a potential substitute, these barriers prevent natural dyes from being widely used (Kusumlata et al., 2024). Changes in industrial practices, customer demand, and technical innovation will all be necessary to make the switch from synthetic to natural colors.

3.3 INTRODUCTION TO HEAVY METALS

Heavy metals are naturally occurring elements that possess high atomic weights and densities. These elements—lead (Pb), mercury (Hg), arsenic (As), cadmium (Cd), and chromium (Cr)—are recognized as harmful to both human health and the environment, even at minimal concentrations. A notable characteristic of heavy metals is their ability to persist in the environment without breaking down, leading to prolonged pollution of ecosystems. These metals are recognized for their ability to bioaccumulate within living organisms, potentially moving up the food chain and resulting in negative impacts on different species, including humans (Järup, 2003). Sources of heavy metals in the environment are mainly divided into natural and anthropogenic categories (Figure 3.1). Natural events that emit trace amounts of these metals into the soil and air include volcanic eruptions, rock erosion, and wildfires. However, the levels of heavy metals in the environment have risen significantly due to human

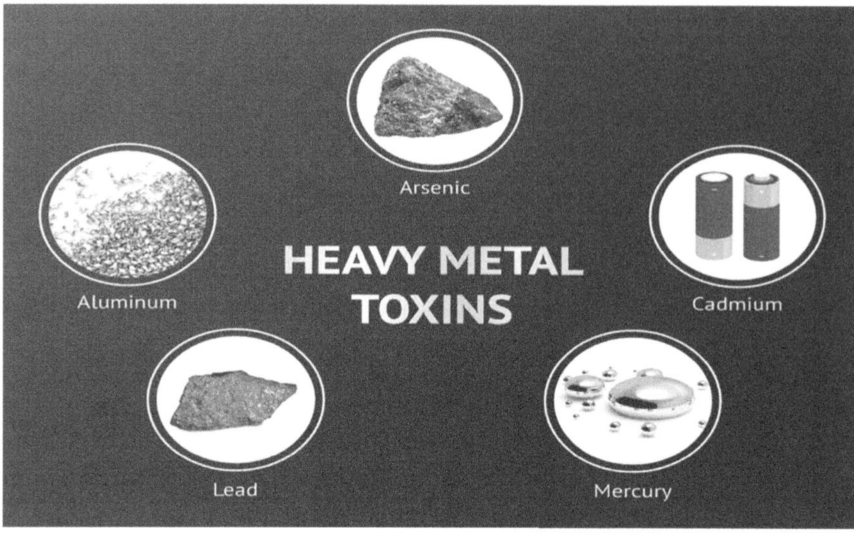

FIGURE 3.1 Different heavy metals present in the environment.

actions, often reaching hazardous levels. Significant human-caused origins comprise industrial discharges from mining and smelting, incorrect disposal of electronic waste, utilization of leaded gasoline, and farming methods that utilize metal-containing pesticides and fertilizers (Tchounwou et al., 2012). The discharge of industrial wastewater and the management of municipal waste systems also lead to the pollution of rivers, lakes, and groundwater (Nagajyoti et al., 2010). While harmful heavy metals such as lead and cadmium are well recognized, it is important to differentiate between toxic heavy metals and essential trace metals. The proper functioning of biological systems relies on trace elements such as zinc (Zn), copper (Cu), and iron (Fe), which are vital for enzyme activities and cellular processes. Nonetheless, even these crucial metals can turn toxic at higher concentrations (Bruins et al., 2000). The delicate balance between helpful and toxic metal concentrations underscores the need to comprehend their vital biological functions and possible dangers.

3.3.1 Environmental Pathway

Soil, water, and air are contaminated by heavy metals via different routes. Soil pollution frequently arises from the disposal of industrial waste, the application of pesticides, and mining operations. When heavy metals enter the soil, they can attach to organic material, complicating their removal and causing persistent ecological disruptions (Alloway, 2013). Plants take in these metals via their roots, integrating them into the food chain (Kabata-Pendias, 2000). Water pollution is another major route, through which metals enter via industrial discharges, leaching from waste sites, and acid mine runoff. These metals can build up in rivers, lakes, and groundwater systems, impacting aquatic organisms and human health when ingested via polluted drinking water (Järup, 2003). Pollution from heavy metals, such as those released during mining activities or the burning of fossil fuels, adds to air pollution, landing on bodies of water and soil, thereby broadening their ecological impact (Tchounwou et al., 2012). In ecosystems, heavy metals experience bioaccumulation and biomagnification once they enter the environment. The gradual accumulation of heavy metals within an organism—typically occurring more rapidly than their removal—is known as "bioaccumulation." For example, mercury is recognized for building up in fish, posing dangers to predators, such as humans, that eat them (Clarkson & Magos, 2006). Biomagnification, referred to as bioamplification, is a process where heavy metals and various toxic substances accumulate in greater concentrations as they ascend the food chain. This issue is especially troubling with heavy metals like mercury, lead, cadmium, and arsenic, which remain in the environment and build up in the tissues of living beings. The process starts with these pollutants present in soil or water at low concentrations, which are taken up by primary producers such as algae and plants. Herbivores that eat these producers accumulate the pollutants in greater concentrations due to their consumption of large quantities of the contaminated plants or algae over time (Figure 3.3).

Heavy metals in aquatic ecosystems are especially problematic because fish and other aquatic organisms can absorb them, and they are then passed up the food chain to humans, who are often at the top of these chains. For example, methylmercury, a neurotoxic form of mercury, can accumulate in fish at concentrations harmful to human health when consumed in large quantities. Biomagnification poses significant risks to biodiversity and ecosystems, affecting not only individual species but also entire food webs by altering predator–prey dynamics and reproductive success. The severity of biomagnification is influenced by factors like environmental conditions, metal type, and species-specific traits. Addressing this issue requires comprehensive environmental regulations and pollution control strategies to prevent further contamination of ecosystems.

3.3.2 Health and Ecological Impacts

Heavy metals are naturally occurring elements with high atomic weights and densities at least five times greater than that of water. They include elements like lead (Pb), cadmium (Cd), mercury (Hg), arsenic (As), chromium (Cr), and nickel (Ni), among others. While trace amounts of some heavy metals are essential for biological functions, excessive concentrations can lead to severe toxic effects on both human health and ecosystems (Tchounwou et al., 2012). Heavy metals have detrimental effects on human health when present in elevated amounts, primarily due to their non-biodegradable and bioaccumulative nature. Exposure to these metals, whether through food, water, soil, or air, can result in acute and chronic health conditions. Lead exposure, for example, is particularly harmful to the nervous system, especially in children, where it can impair cognitive development and cause behavioral issues. Even at low levels, lead exposure can result in neurological and developmental problems, and higher exposures can be fatal (Lanphear et al., 2005). Similarly, cadmium exposure, commonly resulting from the consumption of contaminated food or smoking, can cause kidney damage, bone demineralization, and, in severe cases, cancer.

Mercury is an additional extremely poisonous heavy metal, especially in its methylmercury variant, which accumulates in aquatic life. As mercury ascends the food chain, its concentration increases, creating a significant danger for people eating fish. Exposure to methylmercury, especially in pregnant women, is linked to negative impacts on the neurological development of fetuses, resulting in cognitive and motor deficits in children (Clarkson & Magos, 2006). Arsenic, frequently present in groundwater, is very harmful and linked to skin, bladder, and lung cancers, in addition to cardiovascular diseases when ingested over extended durations. From an ecological perspective, heavy metals interfere with numerous environmental processes and threaten biodiversity. In soil, they can impede plant growth by influencing enzyme activity and disrupting nutrient cycles, which subsequently impacts agricultural productivity and food security. Plants subjected to high levels of heavy metals can experience chlorosis, hindered photosynthesis, and stunted growth, eventually resulting in lower yields (Nagajyoti et al., 2010). Heavy metals can adversely affect aquatic ecosystems. When these metals enter water bodies, they build up in sediments, where they are consumed by organisms that dwell on the bottom. This procedure interrupts aquatic food networks and diminishes biodiversity, as species that cannot withstand elevated metal concentrations perish.

Additionally, the existence of heavy metals in ecosystems results in enduring impacts. These metals may remain in soil and sediment for extended periods, potentially lasting for centuries, complicating and driving up the costs of cleanup efforts. The movement of heavy metals in the environment exacerbates the problem, as these contaminants can disperse through air and water, affecting areas distant from the original contamination source (Alloway, 2013). Human actions, including industrial processes, mining activities, agricultural methods, and waste disposal, have significantly increased the concentrations of heavy metals in the ecosystem. Industrial waste significantly contributes to heavy metal contamination, with industries such as mining, smelting, and chemical manufacturing discharging substantial amounts into air, water, and soil. Additionally, the use of pesticides and fertilizers in farming brings heavy metals such as arsenic, cadmium, and lead into the soil, which subsequently infiltrate the food chain (Kabata-Pendias, 2000). In conclusion, heavy metal contamination is a global issue with significant impacts on human health and ecosystems. Controlling and minimizing heavy metal pollution is crucial and involves policy reforms, advancements in waste management technology, and public awareness initiatives designed to decrease exposure. Addressing heavy metal pollution necessitates a collaborative approach involving governments, industry stakeholders, and the community to protect environmental and human health

3.3.3 Regulations and Remediation

Heavy metal pollution has prompted the development of numerous cleanup and remediation techniques in addition to regulatory actions. Phytoremediation is a useful technique that uses plants to remove heavy metals from contaminated soil and water by absorption, stabilization, or degradation. Some plant species, including poplar and willow, have shown exceptional abilities to remove lead and cadmium, thus lowering the toxicity of the soil. Another method is **bioremediation**, which employs microorganisms to degrade or transform heavy metals into less toxic forms, enhancing soil health and ecosystem recovery (Glick, 2010). The World Health Organization (WHO), the Food and Agriculture Organization (FAO), and other international bodies have set maximum allowable concentrations for heavy metals in food and water. For instance, the WHO has recommended limits for metals like lead in drinking water to prevent long-term exposure effects. In the United States, the Environmental Protection Agency (EPA) enforces the Clean Water Act and the Safe Drinking Water Act, which regulate pollutants in surface water and drinking water sources, respectively. Similarly, the European Union's REACH regulation addresses the risks posed by chemicals, including heavy metals, ensuring safer usage in consumer products and industrial applications.

In agriculture, guidelines from the Codex Alimentarius and FAO have established permissible levels of heavy metals in soil, emphasizing the need to monitor agricultural lands contaminated by industrial effluents. Heavy metals in soil can be absorbed by crops, entering the food chain and posing a health hazard to humans and animals. Many countries have adopted these guidelines as standards for their soil quality assessment programs. Remediation techniques for heavy metals are essential for environmental cleanup and reducing human exposure. Common strategies include phytoremediation, bioremediation, and chemical methods. Phytoremediation, for example, utilizes plants known to accumulate heavy metals in their tissues, which can then be harvested and safely disposed of. This method is eco-friendly and cost-effective, making it suitable for large-scale applications. Bioremediation employs microorganisms that transform heavy metals into less toxic forms through metabolic processes. Chemical remediation, on the other hand, involves using reagents that bind to heavy metals, immobilizing them in soils or precipitating them out of water bodies. Effective management of heavy metal contamination requires collaboration across multiple sectors, including government agencies, industries, and research institutions. Continuous monitoring and updating of safety standards are essential to address emerging contamination sources and advancements in remediation technologies.

3.4 OTHER HAZARDOUS SUBSTANCES FOR THE ENVIRONMENT

3.4.1 Pesticides: An Overview

According to the World Health Organization (WHO), the global impact of pesticide poisoning is staggering, with approximately 3 million cases reported each year and over 2 million deaths, primarily in developing nations. Pesticides, designed to kill, do not discriminate and can harm any organism, including humans (Figure 3.2). Their action leads to the production of reactive oxygen species (ROS), which reduce antioxidant levels in the body, causing oxidative damage.

Pesticides, with their chemical and biological agents, are classified based on their characteristics, such as mode of entry, target pest, function, and chemical composition. Despite their utility, they significantly threaten human health, other species, and the environment. The World Health Organization (WHO) has classified pesticides into four categories based on their toxicity: slightly dangerous, moderately dangerous, highly dangerous, and extremely dangerous, underscoring their potential harm.

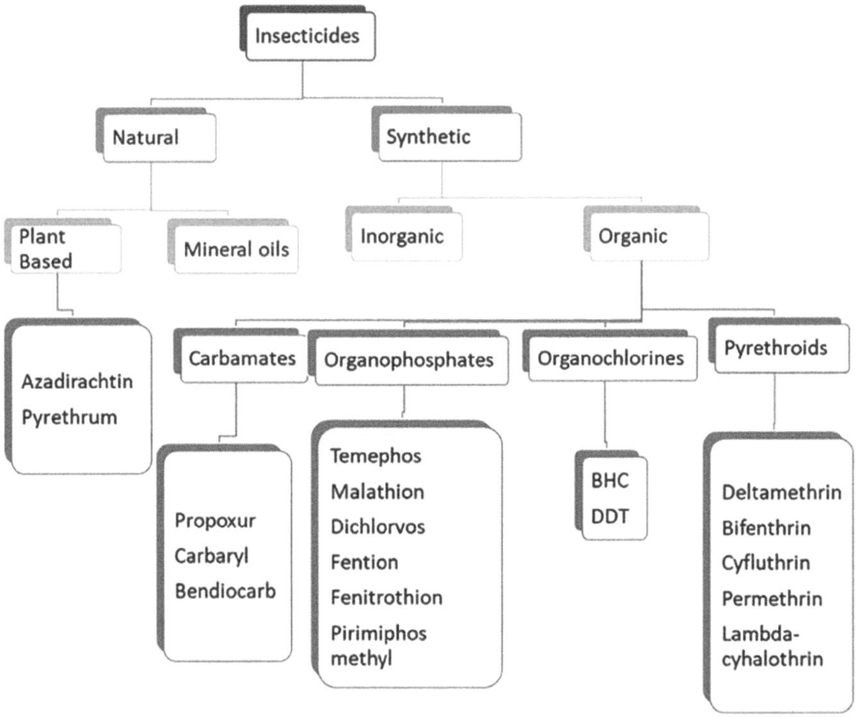

FIGURE 3.2 Classification of pesticides/insecticides.

3.4.2 Effect on Environment

Pesticides create a negative impact on our environment as their risks have slowly started surpassing their uses and applications. Pesticides affect animal and human life, plant biodiversity, aquatic life, terrestrial food webs, and ecosystems. According to a study, 80–90% of pesticides are volatilized within a few days. The most common method of using pesticides is spraying—the pesticides that get violated in the air start targeting and affecting the non-target organisms.

To make it simpler, let us consider an example. We use herbicides that volatize the treated plants, and the vapors cause damage to the non-target plants and flowers in the same environment. Unrestrained use of these pesticides results in a decrease in many aquatic and terrestrial animal and plant species.

Furthermore, the water bodies, soil, cropland, and air are getting contaminated with these chemicals to toxic levels. This is causing several hazards to our biodiversity and the ecosystem. In short, pesticides are a blessing to farmers but at a heavy price. They make agricultural yield faster and easier, with many benefits, but the underlying problem remains. The problem regarding the toxins and chemicals in them poses a risk and hazard to our environment and human health and raises safety concerns.

3.4.3 Health Effects of Pesticides

Pesticides help to control many vector-borne diseases; however, their long-term use can result in various health issues. Infants and children are primarily in danger of these effects due to inadequate or long-term use of pesticides. As the use of pesticides has increased, so has the exposure to these chemicals. Saying that the farmers and pesticide applicators are vulnerable to the toxic effects of pesticides. Pesticides invade the human body through inhalation, ingestion, and penetration via the skin. However, most people are affected when they ingest pesticide-contaminated food. After prolonged intake, the toxins cross the barriers and ingest the human tissue or other storage compartments in our body. Human bodies can eliminate toxins from the body, but in some cases, they stay in the circulatory system. The toxins present in specific types of pesticides lead to Cancer at times. According to Kaur et al., (2019) several studies show risks associated with specific products that have pesticide residues. Some products are fish, seafood, and sometimes milk and dairy products. There is a small but statistical association between specific pesticide residues and cancer risks, including DDD (dichlorodiphenyldichloroethane), DDT (dichloro-diphenyl-trichloroethane), and PCBs (polychlorinated biphenyls). The imbalance in our system affects the cellular signaling pathways and lipids, proteins, and nucleic acid. Oxidative stress and ROS create prolonged health effects like neuro-degeneration, respiratory, cardiovascular, endocrine renal, carcinogenesis, and reproductive problems. Pesticides disrupt the normal functioning of oxidative balance and make way for different diseases. Although the contribution of agrochemicals is elevating with each passing day, it still stands as a hazard to our environment, the ecosystem, and human and animal life. Pesticides are meant to control pests, rodents, weeds, insects, and related diseases in storage. The pesticide industry has undoubtedly made significant progress in developing and producing low-risk, user-friendly, and eco-friendly formulations. However, in underdeveloped and developing nations, pesticides are still in old, conventional formulations.

3.4.4 Chemical Hazards: A Brief Overview

Chemical hazards are substances present in harmful chemicals. They can be classified as corrosive, irritants, asphyxiants, carcinogens, teratogens, mutagens, reactants, flammables, and sensitizers. Workplace exposure to such chemical hazards is called occupational hazards. The number of anthropogenic chemicals, by-products, and manufactured metabolites is increasing daily. Therefore, wildlife and humans are exposed more to complex mixtures. These chemical substances are resulting in nig impact on our biodiversity, ecosystem, and living beings (Drakvik et al., 2020).

3.4.5 Effect on Human Health

Humans have adapted to the modern lifestyle, making it impossible to avoid exposure to chemical pollutants these days. There are different pathways by which we are exposed to hazardous chemicals. It includes:

- Living near an area affected by diffuse pollutants (industrial area)
- Direct contact with chemicals in unknown ways (working in industry, food additives, and preservatives)
- Consumption of food materials harvested from contaminated area (Naidu et al., 2021)

Chemical pollutants can impact human fertility (Leiser et al., 2019). It has been reported that pollutants such as polycyclic aromatic hydrocarbons, dioxins, BFRs, and non-polyphenols affect the male reproductive system. There is a significant decline in the percentage of male fertility globally. Besides, VOCs, POPs in polluted air, metals, etc., cause oxidative stress to male reproductive cells.

3.4.6 Food Chain Pollution

As per the American Academy of Pediatrics reports, around 10,000 chemicals are used or involved in the modern food supply. These chemical pollutants pose a massive threat to humans from consuming chemically infected food. The risks may carry forward to future generations as contaminants were found in human breast milk before. Chemical substances can also affect the human gut microbiome and have long-term impacts on our immunity and metabolism.

Foods get contaminated at several stages before contamination—including during harvesting or animal production, processing, post-harvest storage, transport, etc. Antibiotics, growth-promoting substances, pesticides, PCBs, metal(loids), preservatives, packaging residues, excess nitrates, etc., are contaminating the food chain at a high level that is hazardous to both human health and the environment. This creates an imbalance in fish, vegetables, grains, and livestock via aerial deposition, soil, surface water, or groundwater.

3.4.7 Biodiversity Loss

A significant loss in the Earth's surface layer and its biodiversity is affecting the livestock and vegetation canopy. (Banwart et al., 2024). This is due to an increase in the daily use of agrichemicals, which is creating a chronic impact on our environment and substantial loss in the earth's biodiversity. For example, according to a study in 2013, the global loss of honeybees has led to a global crisis for crop pollination due to neonicotinoid pesticides. Left outs of higher persistent chemicals create greater ecological impact in highly contaminated areas and significant pollution threats to marine water and groundwater. Up to 78% of insect species loss has been reported from 290 sites in Germany. Hence, such chemical hazards alter biological processes, such as soil formation or decomposition in the natural environment, leading to unfavorable environmental circumstances.

3.4.8 Microplastics: Introduction

Plastics are "flexible" or "malleable" in nature. These are synthetic materials that can be molded into virtually any shape. Plastics have high thermal and electrical insulation values. It is inexpensive, strong, durable, lightweight, and corrosion-resistant.

Therefore, the use of plastics keeps increasing. The most popular and widely used synthetic plastics are polyvinyl chloride (PVC), polyethylene (PE), polystyrene (PS), polypropylene (PP), and polyethylene terephthalate (PET). In short, these plastics represent 90% of the total plastic production globally. They are formed by long chains of polymeric molecules made from both inorganic and organic materials like silicon, carbon, hydrogen, and oxygen, mostly fetched from coal, oil, and natural gas (Ivar Do Sul & Costa, 2014).

Microplastics are extremely tiny pieces of plastic detritus present in the environment coming from the breakdown of industrial wastes, and consumer-used plastics (Mitrano & Wohlleben, 2020). Hence, it is crystal clear as to how microplastics are the major compounds polluting the marine and coastal ecosystems.

3.4.8.1 Microplastics and Hazards

The study by Ivar Do Sul and Costa (2014) investigates and critically analyzes the growth of microplastic pollution. Plastic debris is found within every marine habitat. Microplastics have been studied with (a) sandy and middy materials, (b) plankton samples, (c) chemical pollutant interaction, and (d) vertebrate and invertebrate ingestion. Thus, all marine habitats and organism groups are at huge risk of interacting with microplastics.

Microplastics stand as a serious threat to human life due to their toxic effects on our cells and organoids (Li et al., 2023). Microplastics invade the human body through inhalation, oral intake, and skin contact. Therefore, they can constantly increase oxidative stress, damage DNA, and create metabolic disorders, organ dysfunction, immune response, neurotoxicity, and reproductive damage. Present studies suggest that exposure to the toxicity of microplastics also leads to lipid accumulation, liver infection, flora imbalance, etc.

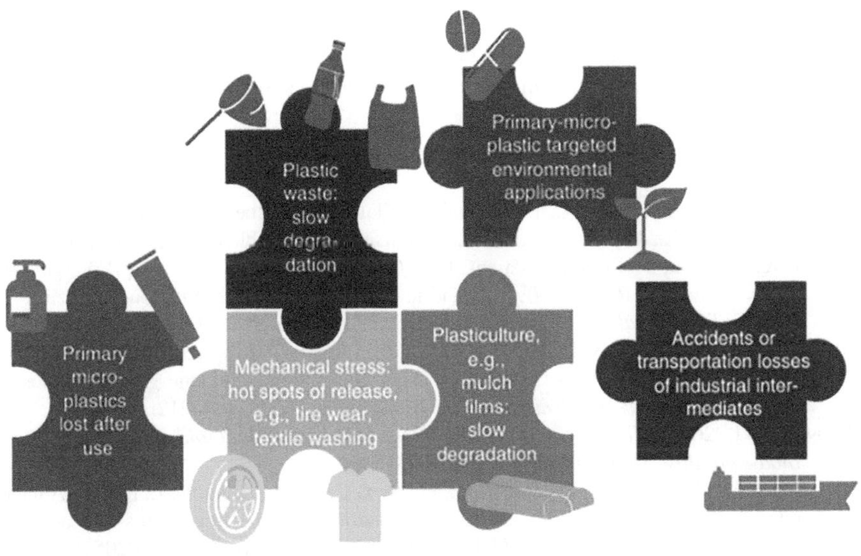

FIGURE 3.3 Environmental source of solid microplastics.

3.4.8.2 Can Microplastics Be Controlled?

Shutting off the use of plastics can bring an end to the plastic life cycles and limit both environmental and health disorders due to these materials. But the million-dollar question is, will it truly end all our problems? Plastics are effective problem solvers during their life cycle, for example, in food safety and packaging, aseptic food packaging, etc. Such types of applications cannot be easily replaced, or a new alternative may have less sustainable scores. Nevertheless, further innovation with the same product is always welcome and possible (Mitrano & Wohlleben, 2020). As Green Chemistry helps in creating and designing chemical products with reduced hazardous substances, one can eliminate the toxicity of plastics. This will promote and support solutions to economic, environmental, and sustainability goals. Green Chemistry is on the way to moderating one plastic pollution source, bringing in biodegradable films and eliminating (micro)plastic wastes in agricultural fields.

Ultimately, the accumulation of plastics will damage our environment, biodiversity, and ecosystem. Thus, replacing or innovating this critical material is truly necessary to achieve sustainable development and protect biodiversity.

3.5 CONCLUSION

In conclusion, heavy metal pollution is still a serious environmental problem that has an impact on ecosystems and human health. In order to lessen the effects of heavy metals, continuous monitoring and remediation efforts are required due to their persistent nature, bioaccumulation, and biomagnification in food webs. Even with improvements in remediation technologies and regulatory frameworks, heavy metal contamination may get worse due to new issues like urbanization, increased industrialization, and climate change. Innovative solutions and immediate attention are needed for emerging sources of pollution, such as nanoparticles and electronic waste. Moreover, remediation efforts and risk assessments may be complicated by the intricate interactions between heavy metals and other environmental contaminants (Järup, 2003). It will take a multifaceted strategy including public awareness campaigns, regulatory actions, and research into sustainable remediation techniques to address these issues.

REFERENCES

Aguilar, F., Charrondiere, U. R., Dusemund, B., Galtier, P., Gilbert, J., Gott, D. M., Grilli, S., Guertler, R., Koenig, J., Lambré, C., Larsen, J.-C., Leblanc, J.-C., Mortensen, A., Parent-Massin, D., Pratt, I., Rietjens, I. M. C. M., Stankovic, I., Tobback, P., Verguieva, T., & Woutersen, R. A. (2009). Scientific opinion on the re-evaluation Tartrazine (E 102). *EFSA Journal*, *7*(11). https://doi.org/10.2903/j.efsa.2009.1331

Ahmad, M. F., Hassan, S., & Imran, Z. (2023). Green approach to water purification: investigating methyl orange dye adsorption using chitosan/polyethylene glycol composite membrane. *Journal of Polymers and the Environment*, *32*(1), 194–212.

Alloway, B. J. (2013). *Heavy Metals in Soils: Trace Metals and Metalloids in Soils and Their Bioavailability*. Springer Science & Business Media.

Banwart, S. A., Nikolaidis, N. P., Zhu, Y.-G., Peacock, C. L., & Sparks, D. L. (2024). Annual review of earth and planetary sciences soil functions: Connecting earth's critical zone. www.annualreviews.org. https://doi.org/10.1146/annurev-earth-063016-020544

Bruins, M. R., Kapil, S., & Oehme, F. W. (2000). Microbial resistance to metals in the environment. *Ecotoxicology and Environmental Safety*, *45*(3), 198–207. https://doi.org/10.1006/eesa.1999.1860

Clarkson, T. W., & Magos, L. (2006). The toxicology of mercury and its chemical compounds. *Critical Reviews in Toxicology*, *36*(8), 609–662. https://doi.org/10.1080/10408440600845619

Cui, R., Zhou, L., Jiang, Y., Li, Y., Zhao, X., & Lu, L. (2020). High performance blue and white fluorescent organic electroluminescent devices with conventional electron transport material as blue emitter. *Dyes and Pigments*, *178*. https://doi.org/10.1016/j.dyepig.2020.108354

Drakvik, E., Altenburger, R., Aoki, Y., Backhaus, T., Bahadori, T., Barouki, R., Brack, W., Cronin, M. T. D., Demeneix, B., Hougaard Bennekou, S., van Klaveren, J., Kneuer, C., Kolossa-Gehring, M., Lebret, E., Posthuma, L., Reiber, L., Rider, C., Rüegg, J., Testa, G., & Bergman, Å. (2020). Statement on advancing the assessment of chemical mixtures and their risks for human health and the environment. In *Environment International* (Vol. 134). Elsevier Ltd. https://doi.org/10.1016/j.envint.2019.105267

Ghais, A., Ahmed, D., Siddig, E., Elsadig, I., Albager, S., Yeh, T. Y., Kossobokov, V., Mouël, J.-L. Le, Allègre, C., da Costa, T. C. E. C., Viana, J. H. M., Ribeiro, J. L., Agarwal, M., Pathak, R., Kant, R., Aragón-Aguilar, A., Hernández-Ochoa, A., & López-Blanco, S. (2020). *International Research in Environment, Geography and Earth Science* (Vol. 2). Book Publisher International (a part of SCIENCEDOMAIN International). https://doi.org/10.9734/bpi/ireges/v2

Glick, B. R. (2010). Using soil bacteria to facilitate phytoremediation. *Biotechnology Advances*, *28*(3), 367–374. https://doi.org/10.1016/j.biotechadv.2010.02.001

Golka, K., Kopps, S., & Myslak, Z. W. (2004). Carcinogenicity of azo colorants: Influence of solubility and bioavailability. *Toxicology Letters*, *151*(1), 203–210. https://doi.org/10.1016/j.toxlet.2003.11.016

Hagan, E., & Poulin, J. (2021). Statistics of the early synthetic dye industry. *Heritage Science*, *9*(1). https://doi.org/10.1186/s40494-021-00493-5

Ivar Do Sul, J. A., & Costa, M. F. (2014). The present and future of microplastic pollution in the marine environment. *Environmental Pollution*, *185*, 352–364. https://doi.org/10.1016/J.ENVPOL.2013.10.036

Järup, L. (2003). Hazards of heavy metal contamination. *British Medical Bulletin*, *68*, 167–182. https://doi.org/10.1093/bmb/ldg032

Javaid, R., & Qazi, U. Y. (2019). Catalytic oxidation process for the degradation of synthetic dyes: An overview. *International Journal of Environmental Research and Public Health*, *16*(11). https://doi.org/10.3390/ijerph16112066

Kabata-Pendias, A. (2000). *Trace Elements in Soils and Plants*. CRC Press. https://doi.org/10.1201/9781420039900

Kaur, R., Mavi, G. K., Raghav, S., & Khan, I. (2019). Pesticides classification and its impact on environment. *International Journal of Current Microbiology and Applied Sciences*, *8*(03), 1889–1897. https://doi.org/10.20546/ijcmas.2019.803.224

Kusumlata, A. B., Kumar, A., & Gautam, S. (2024). Sustainable solutions: Reviewing the future of textile dye contaminant removal with emerging biological treatments. *Limnological Review*, *24*(2), 126–149. https://doi.org/10.3390/limnolrev24020007

Lanphear, B. P., Hornung, R., & Ho, M. (2005). Screening housing to prevent lead toxicity in children. *Public Health Reports*, *120*(3), 305–310.

Leiser, C. L., Hanson, H. A., Sawyer, K., Steenblik, J., Al-Dulaimi, R., Madsen, T., Gibbins, K., Hotaling, J. M., Ibrahim, Y. O., VanDerslice, J. A., & Fuller, M. (2019). Acute effects of air pollutants on spontaneous pregnancy loss: A case-crossover study. *Fertility and Sterility*, *111*(2), 341–347. https://doi.org/10.1016/J.FERTNSTERT.2018.10.028

Li, Y., Tao, L., Wang, Q., Wang, F., Li, G., & Song, M. (2023). Potential health impact of microplastics: A review of environmental distribution, human exposure, and toxic effects. *Environment and Health*, *1*(4), 249–257. https://doi.org/10.1021/envhealth.3c00052

Mitrano, D. M., & Wohlleben, W. (2020). Microplastic regulation should be more precise to incentivize both innovation and environmental safety. *Nature Communications*, *11*(1). https://doi.org/10.1038/s41467-020-19069-1

Nagajyoti, P. C., Lee, K. D., & Sreekanth, T. V. M. (2010). Heavy metals, occurrence and toxicity for plants: A review. *Environmental Chemistry Letters*, *8*(3), 199–216. https://doi.org/10.1007/s10311-010-0297-8

Naidu, R., Biswas, B., Willett, I. R., Cribb, J., Kumar Singh, B., Paul Nathanail, C., Coulon, F., Semple, K. T., Jones, K. C., Barclay, A., & John Aitken, R. (2021). Chemical pollution: A growing peril and potential catastrophic risk to humanity. In *Environment International* (Vol. 156). Elsevier Ltd. https://doi.org/10.1016/j.envint.2021.106616

Tchounwou, P. B., Yedjou, C. G., Patlolla, A. K., & Sutton, D. J. (2012). Molecular, clinical and environmental toxicology volume 3: environmental toxicology. *Molecular, Clinical and Environmental Toxicology*, *101*, 133–164. https://doi.org/10.1007/978-3-7643-8340-4

Tripathi, M., Singh, S., Pathak, S., Kasaudhan, J., Mishra, A., Bala, S., Garg, D., Singh, R., Singh, P., Singh, P. K., Shukla, A. K., & Pathak, N. (2023). Recent strategies for the remediation of textile dyes from wastewater: A systematic review. *Toxics*, *11*(11). https://doi.org/10.3390/toxics11110940

Wargala, E., Sławska, M., Zalewska, A., & Toporowska, M. (2021). Health effects of dyes, minerals, and vitamins used in cosmetics. *Women*, *1*(4), 223–237. https://doi.org/10.3390/women1040020

4 Adsorption Mechanism Demonstrating Adsorbent–Adsorbate Interactions for Emerging Contaminants in the Aqueous Media

Susmita Singha Roy, Asmita Mishra, and Nirmalendu Sekhar Mishra

4.1 INTRODUCTION

The escalating worldwide population, urbanization, industrialization, and climate change cause significant effects on the availability and water quality of water resources in numerous countries. The growing water demand leads to considerable stress on water resources, intensifying concerns regarding the provision of clean drinking water, as indicated in Goal 6 of the 17 Sustainable Development Goals (SDGs) established by the United Nations (UN). The discharge of toxic heavy metals, organic micropollutants (OMPs), pesticides, personal care products (PPCPs), pharmaceuticals, and industrial chemicals such as disinfectants, microplastics, surfactants, engineered nanomaterials, and polycyclic aromatic hydrocarbons (PAHs) into various water resources at concentrations ranging from ng/L to μg/L is a matter of grave concern (Ogbeh et al., 2024).

Organic pollutants, a category of highly harmful synthetic organic chemicals, exhibit extensive persistence under natural conditions owing to their structural stability (Molla et al., 2021). The majority of them are aromatic compounds, consisting of polymers formed from cyclic or planar structures, rendering them as one of the most stable and enduring molecules. Furthermore, PAHs, particularly chlorinated hydrocarbons, phenols, phthalates, and dyes, are the most commonly accumulated substances in natural environments. Additionally, organo-cyanides, including nitriles and long-chain aliphatic compounds (present in various plastics and insulating materials like polyurethanes), as well as organophosphates, pyrethroid herbicides, and

DOI: 10.1201/9781003767251-4

pesticides subsequently add up to this accumulation (Ufarté et al., 2015). They have been classified as persistent organic pollutants (POPs) and have drawn substantial attention due to their extremely toxic and bio-accumulative properties, warranting their removal from aqueous media (Figure 4.1) (Geetha & Nagarajan, 2021).

The effective techniques for eliminating toxic organic substances from water have received a lot of attention. Various techniques, including coagulation, coagulation-assisted filtration, adsorption, precipitation, ion exchange, ozonation, reverse osmosis, and advanced oxidation processes, have been applied to eliminate these heavy metals and organic contaminants from contaminated water and wastewater (Lee et al., 2024). However, these technologies appear to be economically impractical due to their large capital and operational expenses (Nageeb, 2013). Among the several water treatment technologies, the adsorption process using solid adsorbents has the potential to be one of the most effective ways to treat and eliminate organic pollutants from wastewater because it is easy to develop and requires minimal initial investment and land, imparting the adsorption process a preferential selection over the others. Additionally, the process of adsorption excludes the requirement of costly catalysts, and it does not produce hazardous secondary byproducts that are frequently observed in advanced oxidation processes such as ozonation reactions. Nonetheless, novel adsorbents possessing high adsorption capacity, selectivity, cost-effectiveness, and recyclability are essential for attaining the required competitive advantage necessary for real-time applications.

Activated carbons have historically been used as adsorbents because of their affordability and environmentally beneficial qualities; nevertheless, their limited adsorption capacity limits their use. Metal oxides, carbon-based compounds, and

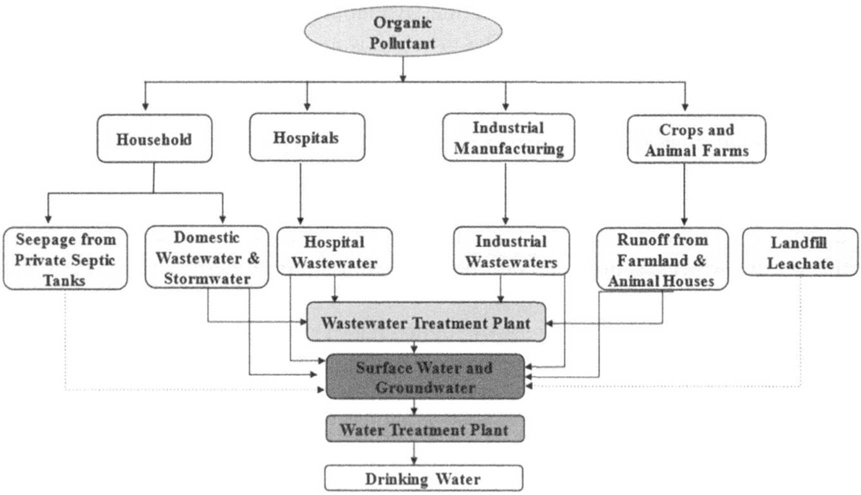

FIGURE 4.1 Schematic diagram of different sources of organic pollutants.

Adapted from Ogbeh et al., 2024.

clay minerals have all emerged as promising adsorbents in the last 10 years. Carbon-based materials stand out among them due to their diverse sources, broad range of porous structures, increased BET surface area, variable surface functional groups, and significant adsorption capability. Accordingly, the carbon-based adsorbents' key characteristics are their high removal efficiency, pH buffering capacity, and simplicity of recovery (Chen et al., 2023a).

This chapter introduces the fundamentals of adsorption, explores the removal of various organic and metallic pollutants from water samples, and sheds light on the factors influencing their adsorption as well as different binding mechanisms with adsorbents.

4.2 ADSORPTION EXPERIMENT

Adsorption generally refers to the transfer of mass between different phases, such as between a gas and a liquid, a liquid and a gas, or solid and solid interfaces. This process is predominantly surface-based and involves the attraction and binding of ions or molecules (known as adsorbates) to the surface of a solid material, commonly called the adsorbent. Conversely, desorption describes the process through which the adsorbate is released from the adsorbent in the opposite direction. The adsorption equilibrium represents a crucial state characterized by a balance between the rates of adsorption and desorption of the adsorbate on and off the adsorbent surface. At this equilibrium point, the concentration of the adsorbate in the liquid phase and the amount bound to the adsorbent surface remain stable over time, given certain environmental conditions such as temperature, pH, ionic strength, and pressure (Ogbeh et al., 2024).

The primary objective of the experiment is to ascertain the suitable composition and quantity of adsorbent for the process of adsorption. For example, a 150-ml conical glass flask filled with 100 ml of water and 0.03–0.3 g of adsorbents was used to conduct the bulk adsorption experiments. The flask was then placed in a mechanistic shaker instrument (Water Bath Shaker Heidolph Unimax 1010 model) at a speed of 200 rpm (Khader et al., 2022). The removal rate R (%) and adsorption capacity ($Q_t\left(mg/g\right)$) were calculated after achieving adsorption equilibrium, using the following equations (M. Chen et al., 2023a)

$$R\left(\%\right) = \frac{C_0 - C_i}{C_0} \times 100 \tag{4.1}$$

$$Q_t\left(mg/g\right) = \frac{C_0 - C_i}{M} \times V \tag{4.2}$$

where C_0 (mg/L), C_i (mg/L), V (L), and M (g) are the initial and concentration at any time (t) of the adsorbate, the volume of the adsorbate solution, and the mass of adsorbent used, respectively.

4.3 ADSORPTION MECHANISM

4.3.1 Hydrophobic Interaction

Hydrophobic interactions refer to the interactions between water and hydrophobic substances (molecules with low solubility in water) (Figure 4.2). Hydrophobic substances are typically nonpolar and often consist of long carbon chains that do not engage with water molecules (Yan et al., 2023). Prior research has demonstrated that hydrophobicity plays a vital role in regulating the ability of microplastics (MPs) to adsorb pollutants (Fang et al., 2019). Organic pollutants are mostly repelled from water onto the surface of the CNMs (carbon nanomaterials) by hydrophobic forces that are either related to the protonation state of ionizable compounds or to the polarity and size of the molecule (Alayan et al., 2021).

4.3.2 Van der Waals Forces

Van der Waals forces are non-covalent and non-ionic interactions that take place between molecules. These forces possess a low level of strength and are confined to the molecular structure, hence being classified as intramolecular. Aliphatic polymers like polyethylene (PE) and polyvinyl chloride (PVC) frequently experience this type of interaction (Hüffer et al., 2018). In addition, micro- and nano-plastics (MNPs) with nonspecific functional groups, such as polypropylene (PP) and polyethylene (PE), primarily interact with organic substances through the van der Waals force, which is a result of the properties of these functional groups (Valiyaveettil Salimkumar et al., 2024).

4.3.3 Electrostatic Interaction

Electrostatic interactions occur when molecules with opposite charges attract each other or when molecules with similar charges repel each other. An electrostatic repulsion or interaction arises based on the charge present on the surface of the adsorbent (Valiyaveettil Salimkumar et al., 2024). Ionizable functional groups present on the surface of the adsorbent engage in electrostatic interactions with the ionizable functional groups found on the molecules of the adsorbate. The nature of this interaction (whether attraction or repulsion) is influenced by pH and is a commonly documented mechanism for uptake. pH is one of the most fundamental parameters that derive the adsorbent's surface charge and the pharmaceuticals' degree of dissociation (Pauletto et al., 2021). Nevertheless, if the pH of the adsorption environment surpasses the acid dissociation constant (pKa) of the organic pollutants, they will lose a proton and adopt an anionic state. This leads to electrostatic repulsion, inhibiting the adsorption process of MPs (Fu et al., 2021). The adsorption of OMPs (organic micropollutants) onto sorbents can be interpreted through molecular electrostatic attraction. The zeta potential of MPCS (magnetic porous carbon-based sorbent) shows a pH-dependent surface charge with a zero-point charge (pHzpc) of 7.6. Below pHzpc, the MPCS surface is predominantly negatively charged (M-O^-), while above pHzpc, it is positively charged (MOH_2^+). At pH 7.04, 4-nitrophenol partially dissociates into negatively charged ions; the positive charge on the MPCS surface at this pH promotes strong electrostatic interactions with 4-nitrophenol, leading to its highest adsorption (D. Chen et al., 2017).

4.3.4 Π–Π Interactions

Adsorption mechanisms of molecules can be significantly influenced by π–π dispersion force within van der Waals forces, occurring between unsaturated (poly)cyclic molecules (Pauletto et al., 2021). Carbon nanotubes and graphene possess identical hexatomic rings of carbon atoms, enabling their spontaneous stacking through π–π interactions. π–π interactions transpire between the delocalized π-electrons of aromatic rings in the adsorbate and those in the adsorbent. The intensity of these interactions depends on a number of factors, including the electronic density and the π-system alignment (Jian et al., 2017). A study was conducted to synthesize two covalent organic frameworks (COFs) for the efficient removal of phenanthrene (PHE). FT-IR analysis was performed to investigate the adsorption mechanism of COFs and PHE. After adsorption, a new absorption peak near 740 cm^{-1}, characteristic of PHE, appeared in the FT-IR spectra of COFs, confirming successful adsorption. Additionally, COFs exhibited a stretching vibration at 3400 cm^{-1}. X-ray photoelectron spectroscopy (XPS) further analyzed the interaction, showing an increase in C–C/C=C band ratios in C1s spectra from 54.09% and 52.76% to 63.38% and 62.25%, respectively, indicating aromatic ring π–π interactions between COFs and PHE, consistent with FT-IR findings (Li et al., 2024). The adsorption mechanism of drug molecules with aromatic rings, such as paracetamol (PCT), is predominantly dependent on π–π dispersion interactions between the aromatic rings of the drug and carbonaceous adsorbents. The –OH and –NH groups in PCT function as potent electron donors, thereby increasing the aromatic ring's electron density (Pauletto et al., 2021). Particularly in acidic environments, resonance effects can facilitate the transition of nitrogen from a proton acceptor to a donor in PCT. Furthermore, the aromatic ring is activated by the amide group in PCT, which facilitates interactions with delocalized electrons in the adsorbent. Therefore, the predominant mechanism of PCT adsorption is π–π interactions (Aminul Islam et al., 2024).

4.3.5 N–Π Interactions

Another form of interaction that may occur in the adsorption of aromatic chemi cals is the n–π interaction. This interaction transpires through the formation of a donor–acceptor complex between π-electron donors and electron-deficient sites of the adsorbate molecules (Pauletto et al., 2021). n–π interactions involve the attraction between lone-pair electrons (n) of heteroatoms (e.g., oxygen or nitrogen) in the adsorbate and the π-electron cloud of the adsorbent. This mechanism is often observed in the adsorption of pollutants containing functional groups such as carbonyls, amines, or nitro groups (Newberry & Raines, 2017). Adsorbents like biochar and graphene oxide, which possess π-electron-rich surfaces, effectively adsorb such compounds. A traditional method and Fourier-transform infrared spectroscopy were used

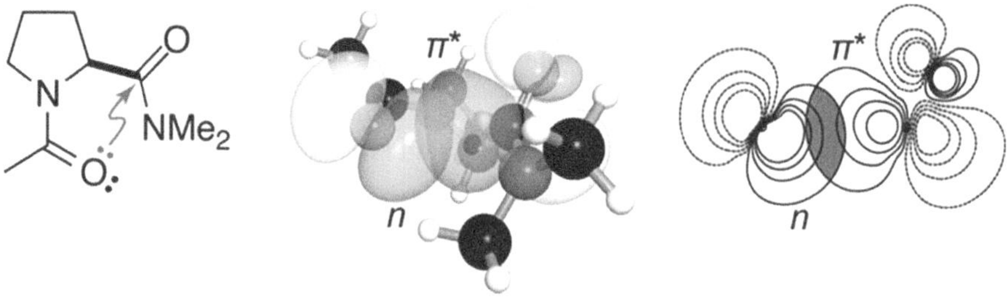

FIGURE 4.2 Overlap of the n and π orbitals (blue) of N-acetylproline dimethyl amide. The overlap integral is 0.1212.

(Adapted from Newberry & Raines, 2017 Copyright 2013 American Chemical Society.)

to validate an additional significant contribution of the n–π interaction between the carbonyl groups on the surface of the biosorbent and the aromatic rings in naproxen molecules. The n–π interaction is not capable of being effectively executed when carbonyl groups are oxidized to carboxylic groups around H_2O_2 treatment (Tomul et al., 2021). The oxygen-containing functional groups on the activated carbon surface, including –COOH and –OH, function as electron donors, whereas the aromatic rings of nimesulide molecules operate as electron acceptors. FT-IR spectra demonstrated a reduction in the strength of the C=O band at 1093 cm^{-1} upon adsorption, signifying the participation of these groups. The reduction was noted for both individual nimesulide adsorption and in a binary system, indicating n → π interactions between the oxygenated functional groups of activated carbon and the aromatic rings of nimesulide (Pauletto et al., 2021). On the surface of the biochar, the carbonyl oxygen ions act as electron donors, whereas the aromatic ring of PCT functions as an electron acceptor. After the adsorption of CAC (commercially activated carbon), the intensity of the pertinent band at about 1630 cm^{-1} (attributable to the C=O groups) significantly diminished. The results indicated a significant role of n–π interaction during the adsorption process (Nguyen et al., 2020).

4.3.6 π–π Electron Donor–Acceptor (EDA) Interactions

The π–π electron donor–acceptor (EDA) interaction is a particular, noncovalent interaction that occurs between electron-rich and electron-poor substances extensively (D. Chen et al., 2017). A study designed PPY/CMC aerogels by using the characteristics and structure of tetracycline, boosting π–π EDA interactions, resulting in high adsorption capacity, a simple production technique, a broad pH buffering range, and the ability to rapidly eliminate contaminants from water (M. Chen et al., 2023a). The adsorption mechanism was elucidated from molecular and electronic viewpoints by density functional theory (DFT) and frontier orbital theory (FOT) tests (Figure 4.3). The examination of electron density, differential charge density, and partial density of states, along with the computation of adsorption equilibrium configuration and binding energy, confirms that

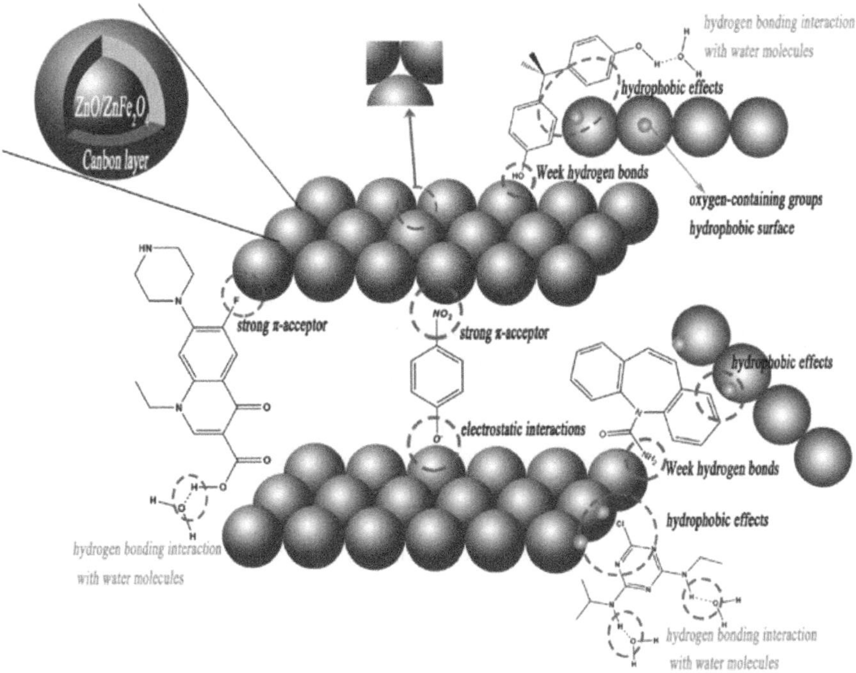

FIGURE 4.3 The schematic illustration of different adsorption mechanisms.

Reused from D. Chen et al., 2017.

TC was adsorbed on PPY/CMC aerogels primarily via π–π EDA interaction (M. Chen et al., 2023a). Researchers showed, due to the π-electron-rich donor characteristics and the presence of hydroxyl, carbonyl, and amino groups, TC (tetracycline) can significantly donate electrons to yttrium-immobilized-graphene oxide–alginate hydrogel (Y-GO-SA), facilitating the formation of a cation complex through cation-π and cation-n bridge interactions (He et al., 2020). The alterations indicate that the active functional groups participate in the adsorption of TC through π–π electron donor–acceptor (EDA) interactions and hydrogen bonding. The high-resolution XPS spectra of O1s after TC adsorption indicate that the binding energies of the C=O and O-H bonds shift to 532.27 and 533.07 eV, respectively. This suggests that the carbonyl and hydroxyl groups may be associated with TC adsorption through n–π electron donor–acceptor interactions and hydrogen bonding, attributable to the presence of benzene rings and highly electronegative atoms in the TC molecule (He et al., 2020). The process of aromatic chemical adsorption onto carbon materials such as MPCS involves the π–π electron donor–acceptor (EDA) interaction (D. Chen et al., 2017). This interaction is affected by aromatic rings that have electron donor (or acceptor) groups, such as –OH or $–NH_2$, or $–NO_2$ or –Cl. Norfloxacin and 4-nitrophenol, which contain electron-withdrawing fluorine (Z. Wang et al., 2010) and $–NO_2$ groups, respectively, function as powerful π-acceptors at neutral pH. Atrazine is a less effective π-acceptor because it has a heterocyclic ring that is π-deficient and a -Cl group. The hydroxyl and amide groups on bisphenol A and carbamazepine make them weak π-donors. Compared to atrazine, bisphenol A, and carbamazepine, norfloxacin and 4-nitrophenol exhibit higher adsorption on MPCS due to the stronger EDA interactions between π-donor and π-acceptor molecules, as opposed to donor–donor or acceptor–acceptor combinations (D. Chen et al., 2017).

4.3.7 Hydrogen Bonding

Hydrogen bonding interactions can be classified into two categories. Dipole–dipole hydrogen bonding interactions frequently transpire between the surface hydrogens of hydroxyl groups (H-donors) on carbonaceous materials and the relevant atoms (i.e., oxygen and nitrogen; H-acceptors) of aromatic adsorbates. Secondly, Yoshida hydrogen bonding interactions frequently transpire between the hydroxyl groups on the surface of carbonaceous materials and the aromatic rings in the adsorbate molecules (Nguyen et al., 2020). The oxygen and nitrogen atoms in the PCT can act as hydrogen acceptors, while the –OH groups on the adsorbent surface can serve as hydrogen donors. This contact is referred to as a dipole–dipole hydrogen bond (Aminul Islam et al., 2024). The adsorption order of bisphenol A, carbamazepine, and atrazine is challenging to interpret. Hydrogen bonds form between the oxygen-containing groups in MPCS (confirmed by XPS) and functional groups like –OH, $-NH_2$, or –NH in these molecules. At neutral pH, bisphenol A, carbamazepine, and atrazine exist in molecular forms, allowing potential hydrogen bonding with MPCS. However, this interaction is likely secondary, as hydrogen bonding with water molecules is more substantial than with the adsorbent surface (Chen et al., 2017).

4.4 INFLUENCING FACTORS

4.4.1 Effect of Solution pH and Ionic Strength

Two operative conditions, the pH of the solution and the ionic strength, significantly influence the adsorption process of an adsorbate onto an adsorbent. These conditions significantly impact the characteristics of the charged surface of the adsorbent and the types of adsorbate present in the solution (Nguyen et al., 2020). The initial pH level of the dye solution significantly affects the degree of ionization, surface charge density, and adsorption capacity of the adsorbent during the adsorption process. At low pH, the removal effectiveness of cationic dye diminishes, whereas that of anionic dye rises due to the strongly protonated surface of the adsorbent, which enhances the adsorption of anionic groups, thus improving overall adsorption. However, at elevated pH levels, it exhibits an inverse order (Bushra et al., 2021). The surface charge of the aged PE MPs became negative as the pH level of the solution increased, which decreased the electrostatic repulsive force and increased MB (Methylene Blue) uptake onto MPs (You et al., 2021). Wang et al. (2023) found that the uptake of MG (malachite green) onto PVC MPs increases with higher salinity. MG, a cationic dye, ionizes to form positively charged ions. High ionic strength promotes dye dimerization in solution, enhancing aggregation and surface adsorption onto MPs (Wang et al., 2023).

4.4.2 Temperature

One of the most critical factors affecting the adsorption process is the temperature. This is useful for determining whether the adsorption process is endothermic or exothermic. The adsorption capacity of activated carbon (AC) derived from date palm rachis and jujube stones using KOH increased with temperature from 298 to 318 K (Daoud et al., 2017). This endothermic and spontaneous process suggests that higher temperatures enhance adsorption by enlarging pore sizes, activating the adsorbent surface, or creating new active sites through bond breakage (Bushra et al., 2021). Another experiment was conducted at 40–60°C to examine the impact of temperature while maintaining constant conditions, revealing that increased temperature correlates with decreased removal efficiency. This finding indicated that the adsorption of Congo red by ZnONPs is exothermic (Debnath & Mondal, 2020). Chen et al. (2023a) investigated the thermodynamics of aniline uptake on PS across different temperatures. The results indicated that $\Delta G > 0$, $\Delta H < 0$, and $\Delta S < 0$, suggesting that the dye uptake process is nonspontaneous and exothermic. The adsorption capacity of MPs for aniline decreased at elevated temperatures (Z. Chen et al., 2023b). Mahabeer et al. (2023) further studied the thermodynamics of MB (dye) adsorption onto MPs in synthetic wastewater and observed increased adsorption at higher temperatures. This occurs because elevated temperatures provide molecules with more kinetic energy, enabling MB dye molecules to overcome the activation energy for chemisorption. Additionally, higher temperatures may enlarge the pores of the MP adsorbent, enhancing the interaction between the dye and the adsorbent surface (Mahabeer et al., 2023).

4.4.3 Functional Groups

Functional groups on adsorbent surfaces significantly influence the adsorption by introducing chemical properties that enhance specific interactions with adsorbates. In the case of LACM (low aromatic carbon materials) with oxidized nitrogen (N–O-), the functional groups provide electron-accepting sites, enhancing EDA interactions with electron-rich regions of methyl blue (MB). The N–O- groups promote stronger chemisorption by forming stable bonds with MB, contributing to the high adsorption capacity (Mengesha et al., 2021). The –COOH and $-SO_3H$ functional groups on graphene materials are crucial for MB adsorption due to their high reactivity and accessibility at the edges of graphene layers. The layered structure of RGO (reduced graphene oxide) limits the exposure of –COOH as an effective adsorption site, reducing its contribution to MB adsorption capacity (Shen & Chen, 2015).

4.5 CONCLUSION

The adsorption of organic pollutants onto different adsorbents is influenced by complicated mechanisms that involve various interactions. These interactions encompass hydrophobic interactions, hydrogen bonding, π–π interactions, n–π interactions, π–π EDA interactions, van der Waals forces, and electrostatic interactions. Adsorption efficiency is affected by the chemical characteristics of the adsorbent, the presence of functional groups on its surface, and environmental variables including pH, temperature, and ionic strength. An in-depth comprehension of these mechanisms establishes a basis for the design and optimization of adsorbents targeting specific pollutants. Recent developments in the functionalization of adsorbent surfaces enhance selectivity and allow for the customization of properties aimed at specific interactions. These advancements facilitate the efficient removal of organic pollutants from water and various environments, thereby supporting sustainable management of pollutants and restoration actions.

REFERENCES

Alayan, H. M., Aljumaily, M. M., Alsaadi, M. A., Mjalli, F. S., & Hashim, M. A. (2021). A review exploring the adsorptive removal of organic micropollutants on tailored hierarchical carbon nanotubes. *Toxicological & Environmental Chemistry*, *103*(4), 282–325. https://doi.org/10.1080/02772248.2021.1932888

Aminul Islam, M., Nazal, M. K., Sajid, M., & Altahir Suliman, M. (2024). Adsorptive removal of paracetamol from aqueous media: A review of adsorbent materials, adsorption mechanisms, advancements, and future perspectives. *Journal of Molecular Liquids*, *396*, 123976. https://doi.org/10.1016/j.molliq.2024.123976

Bushra, R., Mohamad, S., Alias, Y., Jin, Y., & Ahmad, M. (2021). Current approaches and methodologies to explore the perceptive adsorption mechanism of dyes on low-cost agricultural waste: A review. *Microporous and Mesoporous Materials*, *319*, 111040. https://doi.org/10.1016/j.micromeso.2021.111040

Chen, D., Chen, C., Shen, W., Quan, H., Chen, S., Xie, S., Luo, X., & Guo, L. (2017). MOF-derived magnetic porous carbon-based sorbent: Synthesis, characterization, and adsorption behavior of organic micropollutants. *Advanced Powder Technology*, *28*(7), 1769–1779. https://doi.org/10.1016/j.apt.2017.04.018

Chen, M., Yan, Z., Luan, J., Sun, X., Liu, W., & Ke, X. (2023a). π-π electron-donor-acceptor (EDA) interaction enhancing adsorption of tetracycline on 3D PPY/CMC aerogels. *Chemical Engineering Journal*, *454*, 140300. https://doi.org/10.1016/j.cej.2022.140300

Chen, Z., Yang, J., Huang, D., Wang, S., Jiang, K., Sun, W., Chen, Z., Cao, Z., Ren, Y., Wang, Q., Liu, H., Zhang, X., & Sun, X. (2023b). Adsorption behavior of aniline pollutant on polystyrene microplastics. *Chemosphere*, *323*, 138187. https://doi.org/10.1016/j.chemosphere.2023.138187

Daoud, M., Benturki, O., Kecira, Z., Girods, P., & Donnot, A. (2017). Removal of reactive dye (BEZAKTIV Red S-MAX) from aqueous solution by adsorption onto activated carbons prepared from date palm rachis and jujube stones. *Journal of Molecular Liquids*, *243*, 799–809. https://doi.org/10.1016/j.molliq.2017.08.093

Debnath, P., & Mondal, N. K. (2020). Effective removal of congo red dye from aqueous solution using biosynthesized zinc oxide nanoparticles. *Environmental Nanotechnology, Monitoring & Management*, *14*, 100320. https://doi.org/10.1016/j.enmm.2020.100320

Fang, S., Yu, W., Li, C., Liu, Y., Qiu, J., & Kong, F. (2019). Adsorption behavior of three triazole fungicides on polystyrene microplastics. *Science of the Total Environment*, *691*, 1119–1126. https://doi.org/10.1016/j.scitotenv.2019.07.176

Fu, L., Li, J., Wang, G., Luan, Y., & Dai, W. (2021). Adsorption behavior of organic pollutants on microplastics. *Ecotoxicology and Environmental Safety*, *217*, 112207. https://doi.org/10.1016/j.ecoenv.2021.112207

Geetha, D., & Nagarajan, E. R. (2021). Impact and issues of organic pollutants. In *Management of contaminants of emerging concern (CEC) in environment* (pp. 93–126). Elsevier. https://doi.org/10.1016/B978-0-12-822263-8.00003-8

He, J., Ni, F., Cui, A., Chen, X., Deng, S., Shen, F., Huang, C., Yang, G., Song, C., Zhang, J., Tian, D., Long, L., Zhu, Y., & Luo, L. (2020). New insight into adsorption and co-adsorption of arsenic and tetracycline using a Y-immobilized graphene oxide-alginate hydrogel: Adsorption behaviours and mechanisms. *Science of the Total Environment, 701*, 134363. https://doi.org/10.1016/j.scitotenv.2019.134363

Hüffer, T., Weniger, A.-K., & Hofmann, T (2018). Sorption of organic compounds by aged polystyrene microplastic particles. *Environmental Pollution, 236*, 218–225. https://doi.org/10.1016/j.envpol.2018.01.022

Jian, M. Q., Xie, H. H., Xia, K. L., & Zhang, Y. Y. (2017). Challenge and opportunities of carbon nanotubes. In *Industrial applications of carbon nanotubes* (pp. 433–476). Elsevier. https://doi.org/10.1016/B978-0-323-41481-4.00015-0

Khader, E. H., Mohammed, T. J., Mirghaffari, N., Salman, A. D., Juzsakova, T., & Abdullah, T. A. (2022). Removal of organic pollutants from produced water by batch adsorption treatment. *Clean Technologies and Environmental Policy, 24*(2), 713–720. https://doi.org/10.1007/s10098-021-02159-z

Lee, G., Ahmed, I., Lee, H. J., & Jhung, S. H. (2024). Adsorptive removal of organic pollutants with a specific functional group from water by using metal–organic frameworks (MOFs) or MOF-derived carbons: A review. *Separation and Purification Technology, 347*, 127602. https://doi.org/10.1016/j.seppur.2024.127602

Li, S.-C., Jin, W.-L., Jiang, H.-X., Zhang, H.-X., Kong, D.-M., & Zhu, L.-N. (2024). Efficient removal of phenanthrene by covalent organic frameworks: An experimental combined with theoretical analysis. *Separation and Purification Technology, 336*, 126189. https://doi.org/10.1016/j.seppur.2023.126189

Mahabeer, J., Kumari, U., Lokhat, D., Carsky, M., & Meikap, B. C. (2023). Implementation of microplastics derived from waste plastic for uptake of MB dye: Performance and LCA study. *Desalination, 546*, 116214. https://doi.org/10.1016/j.desal.2022.116214

Mengesha, D. N., Appiah-Ntiamoah, R., & Kim, H. (2021). Azo-dye derived oxidized-nitrogen rich carbon sheets with high adsorption capability for dye effluent under both batch and continuous conditions. *Chemosphere, 279*, 130463. https://doi.org/10.1016/j.chemosphere.2021.130463

Molla, M. A. I., Ahmed, A. Z., & Kaneco, S. (2021). Reaction mechanism for photocatalytic degradation of organic pollutants. In *Nanostructured photocatalysts* (pp. 63–84). Elsevier. https://doi.org/10.1016/B978-0-12-823007-7.00011-0

Nageeb, M. (2013). Adsorption technique for the removal of organic pollutants from water and wastewater. In M. N. Rashed (Ed.), *Organic pollutants—Monitoring, risk and treatment*. InTech. https://doi.org/10.5772/54048

Newberry, R. W., & Raines, R. T. (2017). The $n \rightarrow \pi^*$ interaction. *Accounts of Chemical Research, 50*(8), 1838–1846. https://doi.org/10.1021/acs.accounts.7b00121

Nguyen, D. T., Tran, H. N., Juang, R.-S., Dat, N. D., Tomul, F., Ivanets, A., Woo, S. H., Hosseini-Bandegharaei, A., Nguyen, V. P., & Chao, H.-P. (2020). Adsorption process and mechanism of acetaminophen onto commercial activated carbon. *Journal of Environmental Chemical Engineering, 8*(6), 104408. https://doi.org/10.1016/j.jece.2020.104408

Ogbeh, G. O., Ogunlela, A. O., Akinbile, C. O., & Iwar, R. T. (2024). Adsorption of organic micropollutants in water: A review of advances in modelling, mechanisms, adsorbents, and their characteristics. *Environmental Engineering Research, 30*(2), 230733. https://doi.org/10.4491/eer.2023.733

Pauletto, P. S., Lütke, S. F., Dotto, G. L., & Salau, N. P. G. (2021). Adsorption mechanisms of single and simultaneous removal of pharmaceutical compounds onto activated carbon: Isotherm and thermodynamic modeling. *Journal of Molecular Liquids, 336*, 116203. https://doi.org/10.1016/j.molliq.2021.116203

Shen, Y., & Chen, B. (2015). Sulfonated graphene nanosheets as a superb adsorbent for various environmental pollutants in water. *Environmental Science & Technology*, *49*(12), 7364–7372. https://doi.org/10.1021/acs.est.5b01057

Tomul, F., Arslan, Y., Kabak, B., Trak, D., & Tran, H. N. (2021). Adsorption process of naproxen onto peanut shell-derived biosorbent: Important role of $n - \pi$ interaction and van der Waals force. *Journal of Chemical Technology & Biotechnology*, *96*(4), 869–880. https://doi.org/10.1002/jctb.6613

Ufarté, L., Laville, É., Duquesne, S., & Potocki-Veronese, G. (2015). Metagenomics for the discovery of pollutant degrading enzymes. *Biotechnology Advances*, *33*(8), 1845–1854. https://doi.org/10.1016/j.biotechadv.2015.10.009

Valiyaveettil Salimkumar, A., Kurisingal Cleetus, M. C., Ehigie, J. O., Onogbosele, C. O., Nisha, P., Kumar, B. S., Prabhakaran, M. P., & Rejish Kumar, V. J. (2024). Adsorption behavior and interaction of micro-nanoplastics in soils and aquatic environment. In S. A. Bhat, V. Kumar, F. Li, & S. Kumar (Eds.), *Management of micro and nanoplastics in soil and biosolids* (pp. 283–311). Springer Nature Switzerland. https://doi.org/10.1007/978-3-031-51967-3_11

Wang, K., Wang, K., Chen, Y., Liang, S., Guo, C., Wang, W., & Wang, J (2023). Adsorption–desorption behavior of malachite green by potassium permanganate pre-oxidation polyvinyl chloride microplastics. *Environmental Technology and Innovation*, *30*, 103138. https://doi.org/10.1016/j.eti.2023.103138

Wang, Z., Yu, X., Pan, B., & Xing, B. (2010). Norfloxacin sorption and its thermodynamics on surface-modified carbon nanotubes. *Environmental Science & Technology*, *44*(3), 978–984. https://doi.org/10.1021/es902775u

Yan, W., Wang, Q., Gao, Y., Xu, M., Li, H., Zhou, Y., Liu, C., & Xiao, Y. (2023). Coupling between increased amounts of microplastics and dissolved organic compounds in water. *Water*, *15*(23), 4126. https://doi.org/10.3390/w15234126

You, H., Huang, B., Cao, C., Liu, X., Sun, X., Xiao, L., Qiu, J., Luo, Y., Qian, Q., & Chen, Q. (2021). Adsorption–desorption behavior of methylene blue onto aged polyethylene microplastics in aqueous environments. *Marine Pollution Bulletin*, *167*, 112287. https://doi.org/10.1016/j.marpolbul.2021.112287

5 Adsorption Mechanism and Kinetics for Water Treatment

Mukul Swain and Krushna Prasad Shadangi

5.1 INTRODUCTION

Water is vital to all life on Earth and is a key component of biological activities. As a solvent, it makes it possible for different molecules, such as waste products and nutrients, to dissolve and be transported, which is essential for cellular function. For instance, water is essential to human physiology for functions like circulation, digestion, and temperature control. Water's special qualities, such as its high heat capacity, cohesion, and adhesion, aid in preserving steady temperatures inside surroundings and organisms, which is essential for many species' homeostasis. In addition, photosynthetic processes in plants release oxygen and organic compounds used to sustain most terrestrial life forms, which require water for biochemical processes (Sharma et al., 2020). It also provides accommodation for numerous creatures in aquatic ecosystems, promoting biodiversity and maintaining food chains. It is essential to almost every part of life, hence its scarcity or contamination can have a negative effect on ecosystems and the survival of species.

Numerous contaminants can cause contamination of the water, which has a substantial adverse effect on ecosystems and public health. Inappropriate garbage disposal, mining operations, and industrial discharges are common ways for heavy metals like lead and mercury to get into water bodies. Rivers and lakes become contaminated by organic pollutants, such as pesticides and medications, which are mostly produced by urban wastewater and agricultural runoff (Peng et al., 2025). Waterborne illnesses are often caused by pathogens, including bacteria and viruses, that enter water systems through untreated sewage and agricultural runoff (Forstinus et al., 2016). Also, microplastics contaminate freshwater, which creates adverse effects on human health. These contaminants work together to deteriorate water supplies, making efficient treatment techniques necessary to guarantee everyone has access to safe and clean water.

Its scarcity or contamination poses significant threats to organisms worldwide, disrupting ecosystems and compromising species survival. Insufficient water availability has a direct effect on plant development, which lowers agricultural production and depletes herbivores' food supplies (Falkenmark and Rockström, n.d.). This influences predators further down the food chain. Lower water levels and quality in

DOI: 10.1201/9781003767251-5

aquatic environments can cause many species to lose their habitat, which can affect biodiversity and disrupt the ecosystem's function (Dudgeon et al., 2006). Humans suffer from poor drinking, sanitation, and food production due to water scarcity, which increases the danger of starvation, waterborne diseases, and economic instability. Additionally, humans have to compete for a few resources when freshwater sources dry up or get contaminated, which can create stress and perhaps lead to population drops. In general, water scarcity compromises biological processes, disrupts ecological equilibrium, and places the sustainability of species in threat, emphasizing the necessity of sustainable water management and efficient water treatment technologies (Mo Li et al., 2020).

Classical water treatment techniques are necessary to guarantee the safety and purity of drinking water (Figure 5.1). The main techniques include disinfection, filtration, sedimentation, and coagulation. The process of coagulation is mixing chemicals, such as alum, with water to create flocs that catch contaminants and remove them as they sink to the bottom through sedimentation. Following it, filtration removes pathogens and smaller particles from the water by using a variety of materials, including sand and gravel. As a last stage, disinfection usually uses ultraviolet (UV) light, ozone, or chlorine to get rid of any leftover germs and make sure the water is safe to drink (Khasawneh and Palaniandy, 2021). Organic matter, turbidity, and infections are only a few of the contaminants that these approaches are effective at addressing. However, problems including the emergence of new pollutants and the production of disinfection byproducts make continuous improvements in treatment technology necessary (Guangpeng Zhang et al., 2025). To adhere to regulations and safeguard the public's health, regular monitoring and modification of traditional techniques are essential.

Despite their effectiveness, classical water treatment techniques, including coagulation, sedimentation, and chlorination, have significant limitations. They frequently fail to eliminate new pollutants that are becoming more common in water sources,

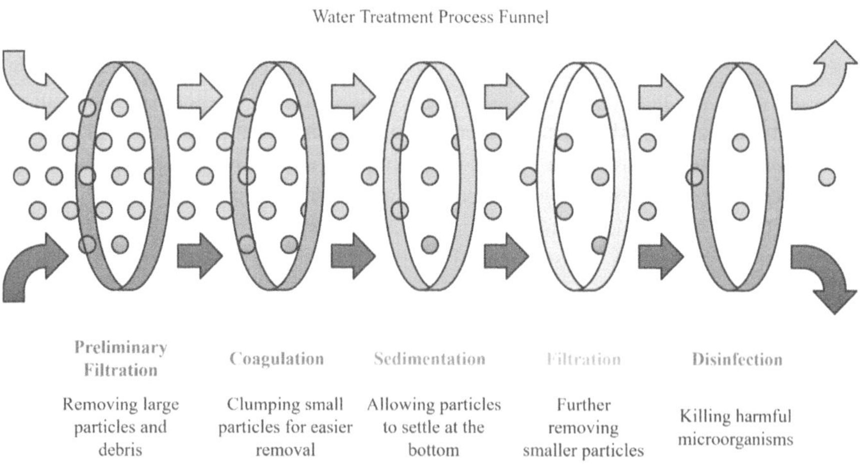

FIGURE 5.1 Classical water treatment process.

such as medications, microplastics, and certain chemical compounds. Furthermore, toxic byproducts (such as trihalomethanes) produced by disinfection techniques like chlorination might be dangerous to human health. Additionally, these procedures frequently require large amounts of energy and chemicals, which increases operating expenses and raises environmental issues (Kumar and Xagoraraki, 2010). To address these challenges, adsorption has emerged as a highly effective and adaptable alternative. Adsorption effectively removes a variety of organic and inorganic pollutants by using materials such as activated carbon, biochar, and zeolites that trap contaminants through surface interactions (Al-Amrani and Onaizi, 2024). Figure 5.2 reflects the types of adsorbents used to remove various contaminants from water. Adsorption uses less energy and chemicals than traditional techniques because it may function well in moderate environments and does not produce any hazardous byproducts. Adsorption is also adaptable and economical for a range of water treatment requirements since it can be tailored to target particular contaminants. Because of its versatility, adsorption offers potential for enhancing water quality and resolving the limitations of conventional treatment techniques in a sustainable manner (Gupta and Suhas, 2009).

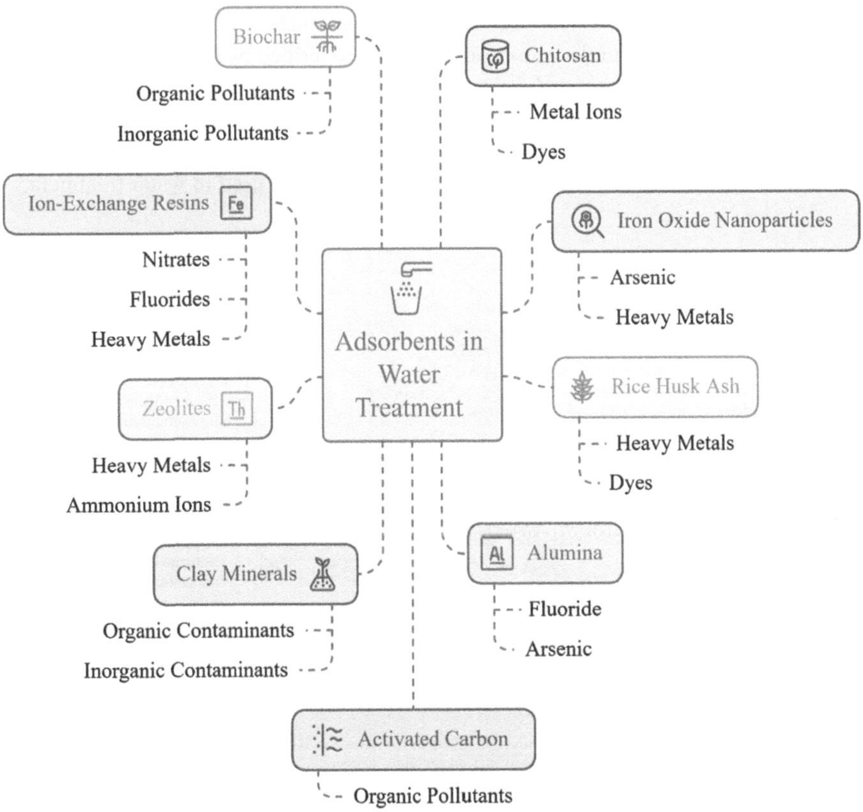

FIGURE 5.2 Adsorbent and its type used for various water treatment processes.

According to the principle of adsorption, contaminants in water (adsorbates) adhere to a solid material (adsorbent) as water flows over it. The pollutant and the surface of the adsorbent are held together by intermolecular forces. In general, the process can be separated into two categories: chemisorption, which forms stronger chemical interactions, and physisorption, which involves weaker van der Waals forces. Determining the appropriate materials for the targeted pollutants is crucial, considering the adsorbent's surface area, porosity, and surface chemistry collectively have a significant impact on how effectively it adsorbs (Ahmad et al., 2014).

Adsorption-based water treatment typically employs a variety of materials, each with specific characteristics that make it suitable for treating specific contaminants. For example, activated carbon acts effectively against organic contaminants and certain heavy metals because of its high surface area and porosity. While materials like charcoal and clay minerals provide reasonable alternatives for wider applications, zeolites are suitable for the removal of heavy metals and ammonium ions (Foo and Hameed, 2010). Metal–organic frameworks (MOFs) and ion-exchange resins are examples of designed adsorbents that have been developed recently to enhance removal efficiency and selectivity, especially for contaminants like nitrate and fluoride (Bedia et al., 2020). Pollutant concentration, contact time, and pH all affect the adsorption process, which in turn affects the removal capacity and efficiency. The equilibrium behavior is explained by adsorption models such as the Langmuir and the Freundlich isotherms, but the rate of adsorption is described by kinetic models such as pseudo-first and pseudo-second order equations. These findings provide an approach for streamlining adsorption procedures in water treatment while balancing variables including regeneration, adsorbent capacity, and cost-effectiveness. Figure 5.3 represents a comprehensive overview of the adsorption process used in water treatment.

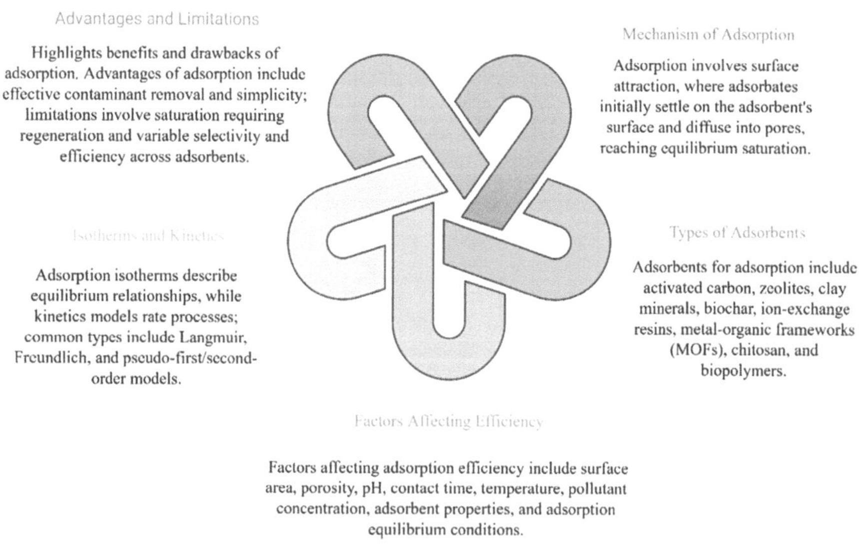

FIGURE 5.3 Overview of the adsorption process used in the water treatment process.

5.2 FUNDAMENTALS OF ADSORPTION

Adsorption is a surface phenomenon in which a solid or liquid (adsorbent) has molecules, atoms, or ions (adsorbates) accumulating on its surface to form a film of molecules or atoms. This technique has become very important in water treatment applications because it is effective in eliminating a variety of contaminants, such as heavy metals, organic compounds, and new pollutants. Adsorption and absorption are essentially different processes. Adsorption primarily takes place at the interface between phases, whereas absorption involves substances penetrating into the bulk of a solid or liquid phase. The solid surface where accumulation takes place is referred to as the adsorbent, and the material that builds up on the surface is known as the adsorbate. Desorption is the opposite process, in which adsorbed materials get released back into the bulk phase (Satyam and Patra, 2024).

The two primary types of adsorptions are chemical adsorption, or chemisorption, and physical adsorption, or physisorption (Figure 5.4). The nature of the adsorbate–adsorbent interaction, temperature dependence, and binding energies of these mechanisms differ substantially.

5.2.1 Adsorption Theories

5.2.1.1 Physisorption (Physical Adsorption)

A fundamental process known as "physisorption" or "physical adsorption" occurs when molecules of the adsorbate adhere to the surface of an adsorbent because of weak van der Waals forces. Its reversible nature, low activation energy, and multilayer

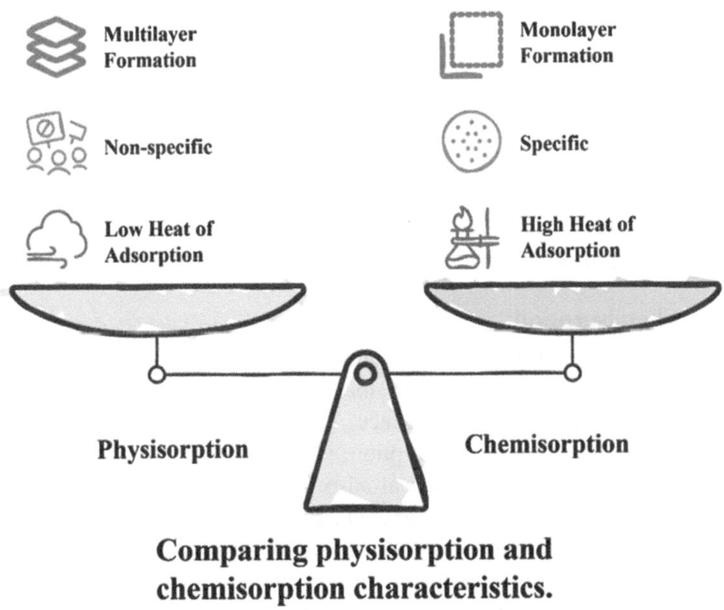

FIGURE 5.4 Comparison between physical and chemical adsorption processes.

adsorption capability distinguish physisorption from chemisorption, which involves the creation of chemical bonds. This makes physisorption extremely important in water treatment applications, as quick and economical adsorption procedures are frequently used to remove contaminants (Weber and Smith, 1986).

Intermolecular interactions between the molecules of the adsorbate and the adsorbent surface are an essential mechanism behind physisorption. These forces include hydrogen bonds, dipole–dipole interactions, and London dispersion forces (Gupta and Suhas, 2009). Adsorption typically takes place at the adsorbent material's surface and inside its porous structure. Surface area, pore size, temperature, pressure, and the characteristics of the adsorbent and adsorbate all affect the process (Abdelwahab and Thabet, 2023).

Adsorption isotherms, such as the Freundlich and Langmuir models, are ideal for analyzing physisorption. For heterogeneous surfaces with different binding site energies, the Freundlich isotherm is applicable, whereas the Langmuir isotherm implies monolayer adsorption onto a homogeneous surface (Vigdorowitsch et al., 2021). Regarding kinetics, the pseudo-er model, which considers the chemico-physical interactions between the adsorbent and the adsorbate, frequently offers a better fit for physisorption processes than the pseudo-first-order model (Revellame et al., 2020).

Physisorption has advantages for treating water. At room temperature, its low activation energy speeds up and improves efficiency, making it appropriate for large-scale applications. Furthermore, it does not change the adsorbate chemically, which means that valuable components like precious metals may be recovered and used again (Colozza et al., 2025).

However, physisorption has its limitations. It is less successful for some high-energy pollutants due to the weak contact forces, and in some instances, the process's reversibility may result in desorption (Foo and Hameed, 2010). Furthermore, adsorbent regeneration may become difficult in some situations, which could reduce the process's cost-effectiveness.

Advanced materials, including hybrid composites and nanomaterials, are being developed, which is broadening the use of physisorption in water treatment. The goal of research is to address new contaminants such as microplastics and medications by improving surface functionalization and boosting adsorption capabilities. Physisorption efficiency can be increased by employing strategies including surface modification and the addition of functional directions (AlAqad et al., 2025).

5.2.1.2 Chemisorption

Chemical adsorption, also known as chemisorption, is a fundamental surface chemistry process in which molecules of the adsorbate develop persistent chemical interactions with the adsorbent's surface. The adsorbent and adsorbate properties are significantly altered by chemisorption, which involves the creation of covalent or ionic bonds as opposed to physical adsorption (physisorption), which involves weaker van der Waals forces (AlAqad et al., 2025).

The high heat of adsorption is one of the most distinctive characteristics of chemisorption. The usual range of enthalpy change for chemisorption is 40–400 kJ/mol, which is significantly more than the range of 10–40 kJ/mol for physisorption (Kachel et al., 2020). This high adsorption heat represents the energy required for establishing robust chemical interactions between the adsorbent surface and the adsorbate. Under some circumstances, the high bond strength causes chemisorption to be irreversible, making it a significant factor in applications like catalysis and sensor technologies where steady adsorption is required (Mahmood Aljamali and Obaid Alfatlawi, 2021).

Another characteristic of chemisorption is specificity. Only when the adsorbent and adsorbate form a chemical relationship that makes them compatible, leading to a highly selective process. Because of this specificity, chemisorption typically results in a monolayer on the adsorbent surface as the number of adsorbed molecules is restricted by the precise alignment and bonding requirements. In catalytic processes, precise control over the reactant environment is crucial to enhance specific reaction pathways (Mizuno and Misono, 1998).

Chemisorption and physisorption exhibit different temperature dependencies. Due to the weak van der Waals forces, physisorption often declines as temperature rises; however, chemisorption frequently exhibits an increase in adsorption with increasing temperature to a threshold. This behavior is attributed to the activation energy barrier that must be overcome for bond formation. The energy needed to overcome this barrier is supplied by higher temperatures, which accelerates the chemisorption process (Low, 1960). However, high temperatures over the ideal range might cause adsorbates to desorb or chemical bonds to break, which lowers the overall adsorption effectiveness (Pullumbi et al., 2019).

In many industrial and environmental applications, chemisorption is a vital process. The chemisorption process by which reactant molecules are absorbed on catalyst surfaces is very crucial in heterogeneous catalysis to promote chemical reactions (Hattori, 2015). Furthermore, selective adsorption ensures the elimination of specific targeted pollutants from gas streams, proving chemisorption is an essential procedure in gas purification.

5.2.2 Adsorption Isotherms

The equilibrium relationship between the concentration of adsorbate in the liquid phase and the amount adsorbed on the solid surface at a constant temperature is described by adsorption isotherms. Knowing these isotherms is essential for improving adsorption procedures in water treatment since they reveal information about the effectiveness and capacity of adsorbents to remove impurities. The Langmuir, Freundlich, and BET (Brunauer, Emmett, and Teller) isotherms (Figure 5.5) are the most used isotherm models in adsorption research because they each provide a distinct viewpoint on the surface properties and adsorption mechanism (Al-Ghouti and Da'ana, 2020).

An essential tool to analyze the effectiveness of adsorbents in water treatment is an adsorption isotherm. They shed light on the capacity, adsorption mechanism, and affinity of the adsorbent and adsorbate. Adsorption systems are designed to remove a variety of pollutants, such as organic contaminants, heavy metals, dyes, and pharmaceuticals, by using these isotherms to describe the adsorption behavior.

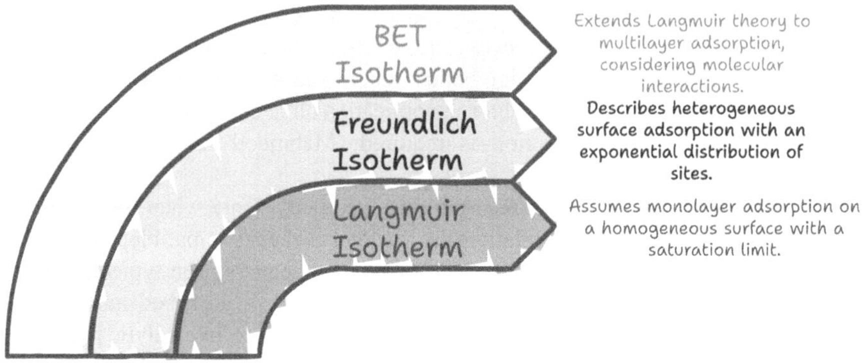

FIGURE 5.5 Types of adsorption isotherms.

5.2.2.1 Langmuir Isotherm

The Langmuir isotherm is a widely used model that describes the adsorption of molecules onto a solid surface, forming a monolayer. Proposed by Irving Langmuir in 1916, it is based on several assumptions, including uniform adsorption sites, no lateral interactions between adsorbed molecules, and a fixed number of adsorption sites. The Langmuir isotherm is particularly applicable to systems where adsorption occurs on a homogeneous surface, providing valuable insights into adsorption mechanisms and surface properties (Swenson and Stadie, 2019).

Mathematically, the Langmuir isotherm is described as in Equation 5.1:

$$q_e = \frac{q_{\max} b C_e}{1 + b C_e} \tag{5.1}$$

where q_e is the amount of adsorbate adsorbed per unit mass of adsorbent at equilibrium (mg/g), $q_{\max}$ is the maximum adsorption capacity (mg/g), b is the Langmuir constant related to the binding affinity (L/mg), and C_e is the equilibrium concentration of the adsorbate in the solution (mg/L). The model predicts monolayer coverage since each adsorption site can only accommodate one molecule.

The Langmuir isotherm is especially useful for determining an adsorbent's maximal adsorption capacity, which is important for applications in gas purification, water treatment, and pollution removal (Liu, 2006). The Langmuir isotherm, for instance, is frequently followed when heavy metals like lead and cadmium are removed from wastewater using zeolites or activated carbon. Adsorption process strength is revealed by the constant b, which represents the affinity between the adsorbent and the adsorbate (Gawande et al., 2017).

The Langmuir isotherm's dimensionless separation factor R_L is one of its primary characteristics and is defined as in Equation 5.2:

$$R_L = \frac{1}{1 + b C_o} \tag{5.2}$$

where C_O is the initial concentration of the adsorbate.

$R_L < 1$ implies favorable adsorption,
$R_L = 1$ represents linear adsorption,
$R_L > 1$ suggests unfavorable adsorption, and
$R_L = 0$ indicates irreversible adsorption.

The Langmuir isotherm also has some limitations, even if it is widely applicable. It makes the assumption that the surface is uniform and ignores multilayer adsorption and heterogeneous adsorption sites. Consequently, other isotherms, such as Temkin or Freundlich, might be more suitable for complicated systems (Alafnan et al., 2021).

To sum up, the Langmuir isotherm is still a fundamental model in adsorption research, offering important information about adsorption efficiency, capacity, and equilibrium behavior in a range of applications.

5.2.2.2 Freundlich Isotherm

The Freundlich adsorption isotherm emerges as a fascinating narrative of molecular interactions, painting a complex picture of how contaminants dance across heterogeneous surface landscapes in water treatment technologies. Unlike its more rigid counterpart, the Langmuir isotherm, the Freundlich model embraces the inherent complexity of real-world adsorption processes, acknowledging that surfaces are not uniform battlegrounds but intricate terrains with varying energetic landscapes (Mozaffari Majd et al., 2022).

Discovered by German physicist Herbert Freundlich in 1909, this empirical model represents a revolutionary perspective on molecular attachment, challenging the previous understanding of adsorption as a uniform, predictable phenomenon. Consider a surface that is not a perfect crystal lattice but rather a rough topography with peaks and valleys of varying binding strengths. In this scenario, molecules do not merely connect with each other. Rather, they negotiate their location through a complex interaction of chemical and physical forces (Wang and Guo, 2020).

The Freundlich isotherm is expressed mathematically as in Equation (5.3):

$$q_e = K_F C_e^{1/n} \tag{5.3}$$

q_e: Amount of adsorbate adsorbed per unit mass of adsorbent (mg/g),
C_e: Equilibrium concentration of the adsorbate in solution (mg/l),
K_F: Freundlich constant representing adsorption capacity,
$1/n$: Heterogeneity factor indicating the intensity of adsorption.

According to this equation, the adsorbent surface is made up of sites with varying adsorption energies, and when the concentration of adsorbate rises, stronger binding sites are occupied first, followed by weaker ones. Surface heterogeneity is measured by parameter $1/n$; values near 0 show strong heterogeneity, while values around 1 indicate homogeneity. Values between 1 and 10 indicate favorable adsorption circumstances, making the parameter a crucial indicator of this interaction. Cooperative adsorption is indicated by values below 1, whereas competitive molecular interactions are suggested by values above 1. The system approaches linear adsorption when n equals 1 (Shimizu and Matubayasi, 2021).

The exponential relationship in the Freundlich model, which is essentially different from linear adsorption models, is what gives it its mathematical beauty. The adsorption capacity bends rather than rises in a straight line with increasing concentration, indicating the heterogeneous character of surfaces. This curving connection implies that a dynamic and adaptable binding landscape is created when initial adsorption takes place at high-energy sites and subsequent molecule attachments demand ever-increasing amounts of energy.

The Freundlich isotherm's comprehensive approach to comprehending the pollutant removal is very advantageous for water treatment technology. Activated carbon is a classic adsorbent that has a complex internal structure made up of millions of microscopic caverns and channels. These surfaces offer different ways for contaminants, such as organic chemicals and heavy metals, to bind. Researchers and engineers can use the Freundlich model as a useful forecasting tool since it not only describes this process but also quantifies the complex molecular interactions occurring at scales that are unseen to the human eye (Lappala, 2024).

Adsorbents are exposed to distinct numbers of target molecules in a series of painstaking laboratory processes to experimentally validate the Freundlich isotherm. By meticulously measuring the equilibrium concentrations and charting the data points, researchers can see the curved adsorption profile. The Freundlich equation can be easily analyzed graphically using linear transformation, and the slope and intercept reveal information about the properties of adsorption. With this systematic approach, abstract molecular interactions are converted into concrete, quantifiable factors that direct the design of water treatment systems (Yang, 1998).

The model's adaptability goes beyond straightforward predictions; it provides a deep understanding of molecular dynamics and surface chemistry. This explains why some contaminants may be eliminated more easily at lower concentrations while others need more intricate removal techniques (Mozaffari Majd et al., 2022). For example, when it comes to eliminating pharmaceutical residues from wastewater, the Freundlich isotherm can be used to forecast the type of activated carbon that would work best, the concentration needed, and any potential purification process constraints.

The Freundlich model is not without its drawbacks, though. It lacks molecular-level mechanistic explanations due to its empirical character, and its precision decreases at very high or low concentrations. So, in order to fully comprehend adsorption processes, researchers need to supplement it with additional isotherms and sophisticated characterization techniques (Shimizu and Matubayasi, 2021).

Modern water treatment techniques are depending more and more on cutting-edge materials like MOFs, graphene, and biomimetic adsorbents (Shimizu and Matubayasi, 2021). To describe these novel materials and better understand and maximize their capacity to remove pollutants, the Freundlich isotherm is still an essential tool. Through the provision of a mathematical structure that recognizes surface variability, the model connects theoretical knowledge with real-world implementation (Lappala, 2024). Beyond water treatment, the Freundlich adsorption isotherm has far wider applications. Applications for its ideas can be found in pharmaceutical development, environmental remediation, agricultural sciences, and catalysis. The recognition that molecular interactions are dynamic, context-dependent processes that necessitate sophisticated, adaptive approaches to understanding rather than being uniform or predictable is a significant scientific breakthrough (Vigdorowitsch et al., 2021).

The Freundlich isotherm, a mathematical technique that changes our comprehension of molecular interactions from a straightforward, homogeneous perspective to a rich, intricate story of adaptive molecular behavior, is a monument to human ingenuity as water scarcity and contamination become more pressing global issues. Every contact in the microscopic realm of adsorption narrates a tale of energy, opportunity, and complex molecular diplomacy, as this brings to mind.

5.2.2.3 BET Isotherm

The multilayer adsorption of gas molecules onto a solid surface can be better understood using the BET adsorption isotherm, a fundamental model in the study of adsorption (Van Erp and Martens, 2011). Stephen Brunauer, Paul Hugh Emmett, and Edward Teller first introduced it in 1938, and it has since grown to be a fundamental concept in the description of porous materials and the comprehension of adsorption processes. The Langmuir model, which presumes monolayer adsorption, is expanded by the BET theory by taking multilayer adsorption into account. It assumes that gas molecules can adsorb in layers onto the surface, with each layer building upon the one before it. In physisorption, when the adsorption forces are comparatively modest and do not entail substantial chemical interactions between the adsorbent and the adsorbate, this assumption is especially pertinent (Sing, 1989).

The BET isotherm is mathematically expressed as in Equation 5.4:

$$\frac{P}{V\left(P_o-P\right)}=\frac{1}{\left(V_mC\right)}+\frac{\left(C-1\right)}{\left(V_mC\right)}*\frac{P}{P_o} \tag{5.4}$$

where:

P: Pressure of the adsorbate gas
V: Volume of gas adsorbed
P_o: Saturation vapour pressure of the adsorbate
V_m: Volume of gas required to form a monolayer
C: BET constant, related to the energy of adsorption in the first layer

In the relative pressure range of 0.05–0.35, where the linear relationship between $P/\left(V\left(P_o-P\right)\right)$ and P/P_o is evident, the BET equation is commonly used. Plotting this linear region allows one to derive the values of V_m and C by figuring out the slope and intercept. Among the many uses for the BET isotherm is the treatment of water. Adsorbents like zeolites, activated carbon, and mesoporous materials are frequently characterized by their surface area and pore size distribution. The effectiveness and adsorption capacity of these materials for eliminating impurities from water are determined by these factors (Goh et al., 2025).

The BET isotherm can be used to investigate the adsorption of different contaminants, such as organic dyes, heavy metals, and medications, in the context of water treatment. Important details regarding the adsorption mechanism, the character of the adsorbent–adsorbate interactions, and the ideal circumstances for efficient contamination removal can be discovered by examining the adsorption isotherms. Additionally, the regeneration potential of adsorbents can be evaluated using the BET isotherm. It is possible to assess the degree of adsorbent recovery and its reusability by looking at the adsorption–desorption cycles (Renu and Sithole, 2024). The design and optimization of adsorption-based water treatment procedures depend on this knowledge.

In summary, adsorption processes in water treatment can be effectively understood and described using the BET adsorption isotherm. The development of effective and sustainable water treatment methods depends heavily on the BET isotherm, which offers information about the surface area, pore structure, and adsorption capacity of adsorbents.

5.2.3 Factors Affecting Adsorption in Water Treatment

The surface phenomenon known as adsorption, which occurs when molecules or ions stick to a solid surface, is a common method for treating water. Several variables, such as temperature, pH, and adsorbent properties, can have a major impact on the adsorption process (Figure 5.6).

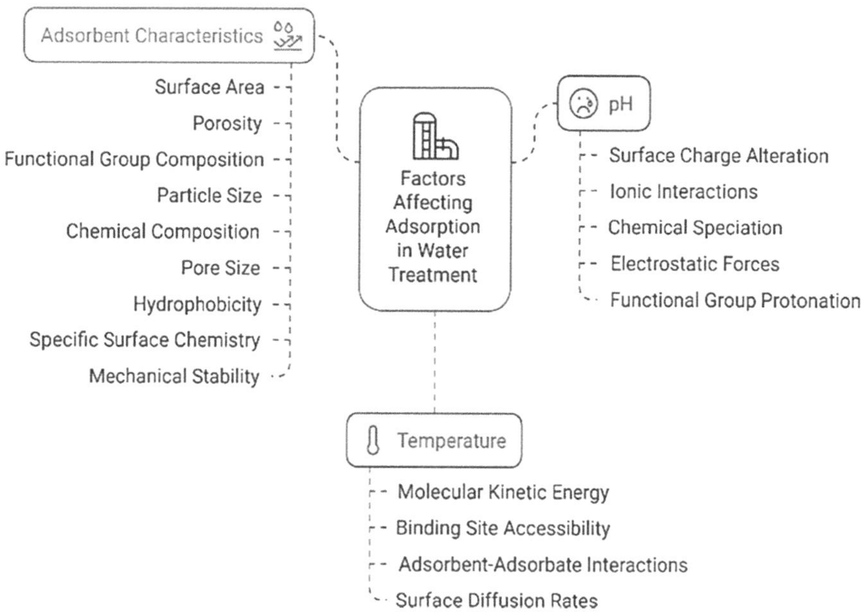

FIGURE 5.6 Factors affecting adsorption in water treatment.

5.2.3.1 Temperature

In adsorption processes, temperature is essential. The rate of adsorption often increases as the temperature rises. The reason for this is that molecules have more kinetic energy at higher temperatures, which increases their mobility and makes it easier for them to interact with the adsorbent surface. But the relationship between temperature and adsorption capability is more nuanced. An increase in temperature frequently leads to a decrease in adsorption capability for physisorption, which involves weak van der Waals forces. This is so that the weak attraction forces between the adsorbent and the adsorbate can be overcome by increased thermal energy. However, because it provides the activation energy required for the chemical reaction, a higher temperature can result in a greater adsorption capacity for chemisorption, which is a process in which strong chemical bonds are produced (Ho and McKay, 1998).

5.2.3.2 pH

Adsorption is also significantly impacted by the pH of the fluid. The ionization state of the adsorbate and the surface charge of the adsorbent can both be strongly affected by pH. For example, the surface of many adsorbents, including zeolites and activated carbon, becomes positively charged at low pH, drawing in negatively charged adsorbates. High pH, on the other hand, causes the surface to become negatively charged, drawing in positively charged adsorbates. Furthermore, the pH may have an impact on the adsorbate's speciation and adsorption behavior. For instance, the adsorption behavior of heavy metal ions can change based on the pH, as they can exist in various forms (such as complexes or free ions) (Yan et al., 2016).

5.2.3.3 Adsorbent Characteristics

Adsorption is greatly influenced by the characteristics of the adsorbent, including surface area, pore size distribution, and surface functional groups. Adsorption capacity rises with a greater surface area because it offers more adsorption sites. The accessibility of the adsorption sites to various-sized adsorbate molecules is determined by the dispersion of pore sizes. Larger molecules are more suited for mesopores and macropores, but micropores are efficient in adsorbing small molecules. Adsorbate molecules can interact with surface functional groups like hydroxyl, carboxyl, and amine groups by a variety of processes, such as complexation, electrostatic interactions, and hydrogen bonding (Vanmoerkerke et al., 2025).

In water treatment, activated carbon is a common adsorbent. Its high surface area and well-developed pore structure allow it to effectively remove a variety of impurities. Activated carbon's hydroxyl, carboxyl, and carbonyl surface functional groups improve its adsorption ability by interacting with adsorbate molecules in a variety of ways (Xiaohong Wang et al., 2022). The adsorption of heavy metals and other contaminants is also successful with zeolites due to their distinct crystalline structures and high cation exchange capabilities (Bedia et al., 2020).

The adsorption process is influenced by these main elements as well as additional operational parameters as contact duration, agitation speed, and adsorbent dosage. Greater adsorption is typically possible as equilibrium approaches with longer contact durations (Karimi et al., 2019).

In summary, temperature, pH, and adsorbent properties are just a few of the many variables that interact to affect the adsorption process in water treatment. The efficacy and efficiency of adsorption-based water treatment systems can be increased by comprehending and maximizing these variables. The development of cutting-edge technologies and techniques to tackle the escalating problems of water pollution and guarantee the supply of safe and clean water depends on this expertise.

5.3 ADSORPTION MECHANISM

The complex process of molecules or ions adhering to a solid surface is known as adsorption, and it is controlled by several molecular interactions (Figure 5.7). Designing effective and efficient adsorbents for water treatment requires an understanding of these interactions.

5.3.1 van der Waals Forces

Electron density variations give rise to weak, non-specific intermolecular forces known as van der Waals forces. The dipole–dipole, dipole-induced, and dispersion forces are among them. In adsorption, dispersion forces for non-polar adsorbates are the most important kind of van der Waals force (Snyder, 1968). In the case of

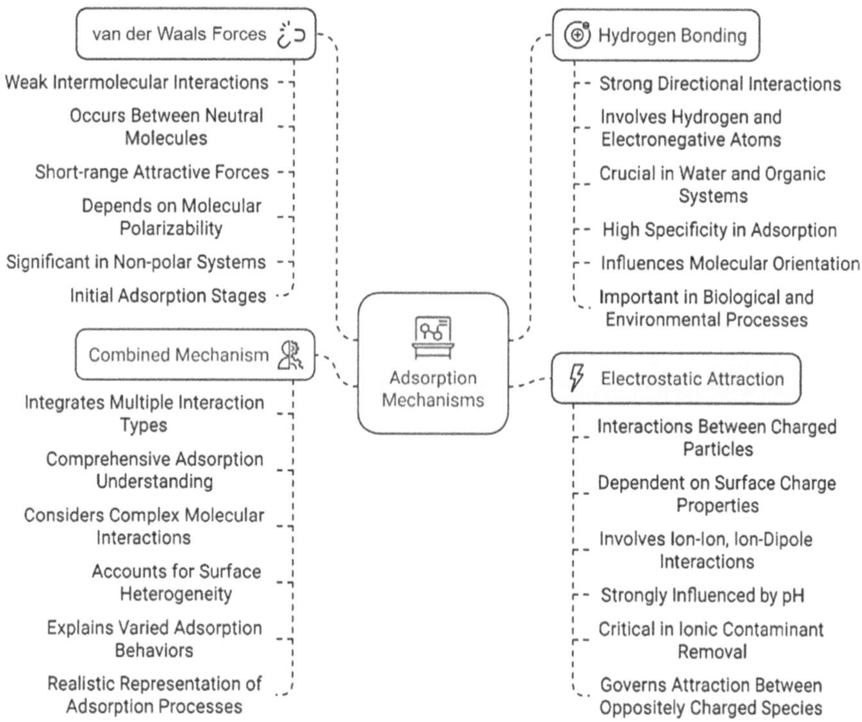

FIGURE 5.7 Various adsorption mechanisms.

adsorption onto a high-surface-area adsorbent, these mild forces can become considerable when several molecules are involved. Through dispersion forces, activated carbon, for example, can efficiently absorb non-polar chemical molecules due to its large surface area and porous structure (Vanmoerkerke et al., 2025).

5.3.2 Hydrogen Bonding

A stronger kind of intermolecular force known as hydrogen bonding takes place when a hydrogen atom is joined to a very electronegative atom (such as oxygen, nitrogen, or fluorine) and a single pair of electrons on another electronegative atom. The adsorption of polar molecules, such as alcohols and water, onto adsorbents having polar functional groups, like hydroxyl and carbonyl groups, depends critically on hydrogen bonding. Through hydrogen bonding, zeolites, for instance, can efficiently adsorb water molecules due to their numerous hydroxyl groups and well-defined pore structure (Gilormini and Verdu, 2018).

5.3.3 Electrostatic Attraction

The interaction of charged species results in electrostatic attraction, commonly referred to as Coulombic forces. When charged adsorbates and charged adsorbent surfaces come into contact during adsorption, electrostatic attraction may take place. Factors including pH, ionic strength, and the existence of surface functional groups can all affect an adsorbent's surface charge. For example, the surface of many adsorbents becomes positively charged at low pH, which attracts negatively charged adsorbates, including heavy metal anions (Lu et al., 2016). The surface becomes negatively charged at high pH, on the other hand, and attracts positively charged adsorbates, like cationic dyes.

5.3.4 Combined Mechanisms

A combination of these molecular interactions is frequently involved in adsorption. Both hydrogen bonds and van der Waals forces, for instance, may be involved in the adsorption of organic contaminants onto activated carbon (Xiaohong Wang et al., 2022). Initial adsorption of the pollutant onto the carbon surface is facilitated by van der Waals forces, but by creating new contacts between the pollutant and the carbon surface, hydrogen bonding can improve adsorption even further (Terzyk, 2003). Heavy metal ion adsorption onto ion exchange resins can also occur by complexation and electrostatic attraction. Initially, the metal ions are drawn to the resin by electrostatic interaction with the charged functional groups, and complexation processes between the metal ions and the functional groups help to stabilize the adsorbed species.

Researchers can improve adsorption procedures for water treatment and create more potent adsorbents by comprehending the fundamental chemical interactions involved in adsorption. With this information, adsorbents with specified surface area, pore size distribution, and surface functional groups can be designed to improve their adsorption capacity and selectivity for target pollutants. Furthermore, a more thorough comprehension of adsorption mechanisms can help with the creation of sophisticated adsorption models and the forecasting of adsorption behavior in diverse scenarios (Renu and Sithole, 2024).

5.3.5 Adsorption Mechanism: Pore Diffusion and Surface Interaction

The complicated field of water treatment reveals a molecular choreography in which adsorbent materials use intricate surface contact and pore diffusion mechanisms to orchestrate a sophisticated symphony of pollutant removal (Figure 5.8). A complex interaction between transport phenomena and molecule recognition is at the core of this process, as evidenced by Xiaowei Zhang et al.'s groundbreaking study in Advanced Materials, which clarified the crucial role that nanoscale surface engineering plays in improving adsorption capacities (Xiaowei Zhang et al., 2024). A basic transport mechanism known as pore diffusion is revealed, in which water molecules move through complex networks of nanoporous structures in a complicated journey that goes beyond straightforward linear motion. Chmelik et al. demonstrated how tortuous paths and diverse pore designs greatly alter mass transfer kinetics, underscoring the crucial significance of pore geometry and connectivity (Chmelik and Kärger, 2020). When pollutants come into contact with the adsorbent surface through boundary layers that are marked by intricate hydrodynamic interactions, the molecular journey starts with external mass transfer. Passive material structures are converted into active molecular recognition platforms through surface contact, which acts as the molecular gatekeeper. Pore diffusion and surface chemistry interact to generate a dynamic, non-equilibrium system in which adsorption performance is influenced by several physicochemical interactions at once. Advanced characterization methods, such as molecular dynamics simulations and in situ spectroscopic investigations, have started to reveal the complex mechanics behind these events (Lappala, 2024). Advanced surface alterations can produce hierarchical porous structures with better mass transfer capacities and pollutant removal efficiency, as shown in a ground-breaking work by Xiaoyun Li et al. (2013), Shao et al. (2025). Target molecules are selectively immobilized by molecular-level interactions, which include intricate electromagnetic forces, electrostatic attractions, and possible hydrogen bonding networks. Recent advances in nanotechnology have introduced new

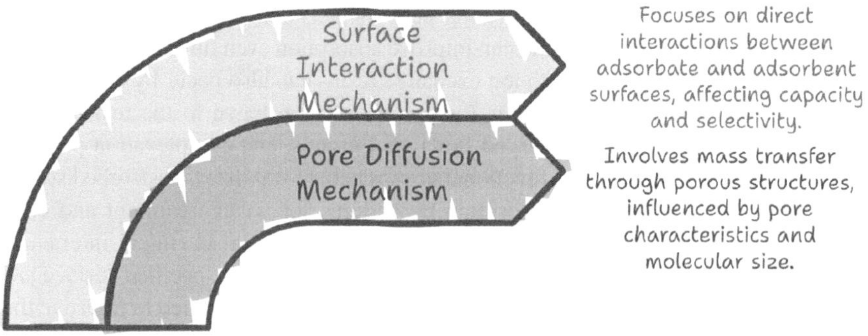

FIGURE 5.8 Mechanism of adsorption.

materials with previously unheard-of molecular recognition capabilities, pushing the limits of conventional adsorption methods. Gadore and Ahmaruzzaman (2021) emphasized the development of intelligent adsorbent materials that can react to environmental changes in real time, providing water treatment methods with previously unheard-of precision (Gadore and Ahmaruzzaman, 2021). To address the problems of water scarcity and contamination worldwide, a fundamental understanding of pore diffusion and surface interaction is essential and goes beyond scholarly curiosity. Environmental engineering, physical chemistry, and materials science are coming together in interdisciplinary approaches to develop revolutionary solutions that promise more cost-effective, sustainable, and efficient water purification methods. Today, the molecular makeup of adsorbent materials offers an advanced platform for the targeted removal of contaminants, capable of handling new problems such as heavy metals, persistent organic pollutants, and pharmaceutical residues. Every nanoscale interaction has the potential to completely transform our strategy for managing water quality worldwide, and the future of water treatment rests in this intricate molecular-level engineering as researchers continue to decipher the intricate principles driving adsorption processes.

5.3.6 Adsorbent Properties

Adsorbent characteristics are the key molecular structure that impacts how well water treatment methods work. The three main characteristics of these properties are surface area, pore size, and functional groups (Figure 5.9). All the properties play crucial roles in the processes that remove contaminants. These characteristics are essential in creating cutting-edge water treatment technologies that can handle challenging environmental issues because they work in concert to determine an adsorbent's final performance.

5.3.6.1 Surface Area

Surface area becomes apparent as the basic canvas on which the complex art of water treatment is created, converting seemingly simple materials into enormous molecular landscapes with previously unheard-of pollutant collection capabilities.

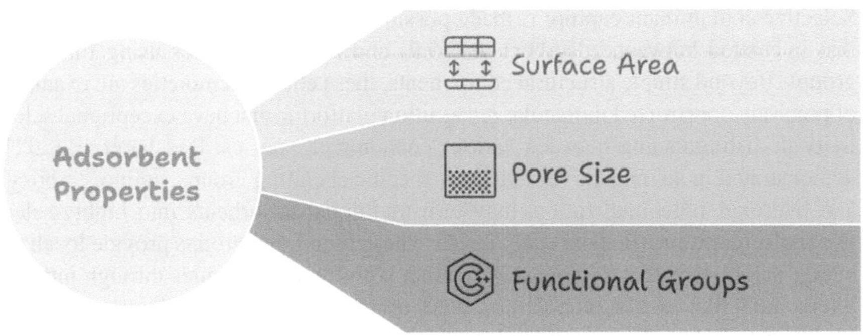

FIGURE 5.9 Various adsorbent properties.

High-performance adsorbents use exceptional specific surface areas, which frequently surpass 1000 m^2/g, to create a molecular playground where water pollutants can interact in ways never seen before. These large surfaces are dynamically created platforms that optimize molecule interface through advanced nanostructuring techniques, rather than being passive domains (Hussain et al., 2025). Surface area engineering has been advanced by advanced materials such as MOFs and graphene-based composites, which have produced three-dimensional structures with exponentially higher adsorption potential. Strategic architectural changes could produce hierarchical porous structures with interconnected networks, significantly increasing accessible surface area and generating multiple interaction sites for targeted contaminant removal (Xiaoyun Li et al., 2013; Xiaowei Zhang et al., 2024). The performance of an adsorbent in water treatment technologies is determined by a complex interplay of geometric configurations, surface topography, and possible interaction sites.

5.3.6.2 Pore Size

Pore size, which ranges from microporous (less than 2 nm) to mesoporous (2–50 nm) and macroporous (more than 50 nm) domains, is a representation of the complex highway system that water molecules and pollutants ride through. Precise pore size engineering can produce molecular sieves with unparalleled selectivity, allowing for the targeted elimination of particular contaminants with nanoscale precision, according to research by Hong et al. (2024) published in Nature Communications (Hong et al., 2024). The kinetics of mass transfer are significantly influenced by the porous architecture, where the distribution of pore sizes, connectivity, and geometric complexity work together to modify molecular transport pathways. The best pore sizes vary depending on the target contaminants and water treatment goals, according to advanced characterization methods, including electron microscopy and nitrogen adsorption isotherms. Multi-modal pore structures with hierarchical architectures could simultaneously address multiple contaminant classes, resulting in versatile adsorbent materials that go beyond conventional single-target approaches, according to a groundbreaking study by Park et al. (2025). More than just a structural factor, the pore size becomes an advanced molecular filter that can distinguish between various molecular species according to their size, shape, and electrical configurations.

5.3.6.3 Functional Groups

Selective contaminant capture is made possible by the complex molecular dialogue that is created between adsorbent materials and water molecules using functional groups. Beyond simple structural components, these chemical moieties are examples of precisely constructed molecular recognition platforms that have exceptional selectivity in distinguishing between various contaminant species. Badsha et al. (2021) demonstrated in an innovative study how specific chelating groups, amino, carboxyl, and hydroxyl functionalization, may turn traditional adsorbents into highly selective molecular traps (Badsha et al., 2021). These functional groups provide localized energy landscapes that dynamically interact with target molecules through intricate interactions like as π–π interactions, electrostatic attractions, and hydrogen bonds. Multifunctional surface modifications could produce adaptive adsorbent materials that can react to changes in the environment and target new contaminants such as heavy metals and pharmaceutical residues. As a dynamic molecular interface where surface chemistry transforms into an advanced recognition mechanism that can be customized to solve particular water treatment difficulties, the chemical intelligence of functional groups goes beyond static capture (Mestri et al., 2025).

5.4 KINETICS OF ADSORPTION

A crucial component of comprehending the dynamics of adsorption processes, especially in water treatment applications, is adsorption kinetics. It explains the rate at which equilibrium is reached and the interactions between adsorbates and adsorbents throughout time. The importance of the reaction rate theory in water treatment is highlighted in this part as it explores this basic method of simulating and analyzing adsorption kinetics. The reaction rate theory offers a scientific lens through which the dynamic interactions between adsorbents and pollutants can be thoroughly examined, and it provides a key framework for comprehending the complex dynamics of adsorption processes in water treatment. By connecting microscopic interactions with macroscopic treatment outcomes, the theory fundamentally clarifies the molecular-level changes that control the rate and effectiveness of contaminant removal.

5.4.1 Reaction Rate Theory in Adsorption Kinetics

Adsorption kinetics is explained by the reaction rate hypothesis, which describes how quickly molecules move from the bulk phase to an adsorbent's surface and then attach to its active sites. According to this hypothesis, adsorption occurs in a sequence of stages, such as surface diffusion, external mass transfer, and direct interaction with the adsorption sites (Figure 5.10). While each step adds to the total rate, the slowest step ultimately determines the rate and is called rate-determining step.

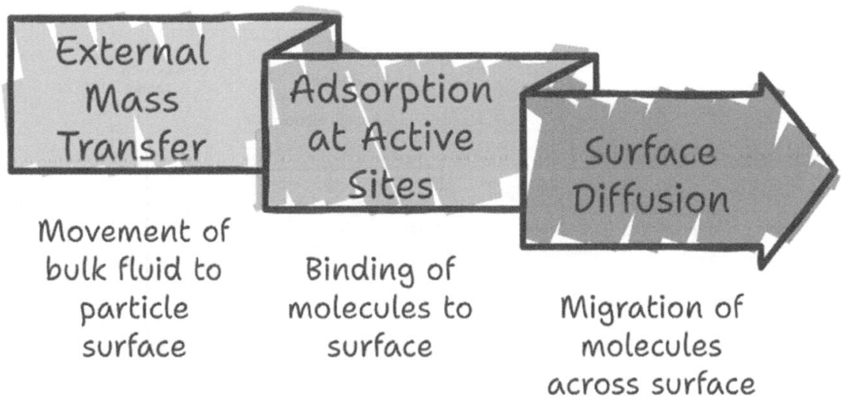

FIGURE 5.10 Adsorption process in mass transfer.

- External Mass Transfer: To reach the boundary layer around the adsorbent surface, adsorbate molecules must diffuse from the bulk solution. The adsorbent size, solution viscosity, and flow parameters all affect this process. For instance, Huang et al. (2011) found that, particularly in low-concentration solutions, external mass transfer dominated the early phases of dye adsorption onto mesoporous silica (Huang et al., 2011).
- Surface Diffusion: Adsorbates move around the adsorbent surface in search of active binding sites after they arrive. This process is dependent on the temperature, surface energy, and porosity of the adsorbent. Achieving high adsorption rates on biochar required the surface diffusion of heavy metals, according to Al-Amrani and Onaizi (2024).
- Adsorption at Active Sites: In this stage, adsorbates are attached to active sites via chemical bonding (such as covalent or ionic bonds) or physical forces (such as van der Waals interactions). The entire response rate for this phase is determined by the energy barrier. Because of the energy needed to create chemical connections, chemisorption on metal oxides is frequently slower, according to research by Blesa et al. (2000).

5.4.2 Mathematical Modeling in Reaction Rate Theory

To quantitatively describe adsorption kinetics, researchers often rely on kinetic models derived from reaction rate theory (Figure 5.11). These models are essential for predicting adsorption rates and designing efficient water treatment systems.

5.4.2.1 Pseudo-First-Order Kinetic Model

The pseudo-first-order model, introduced by Lagergren in 1898, assumes that the rate of adsorption is proportional to the number of unoccupied active sites. The model is expressed in Equation 5.5:

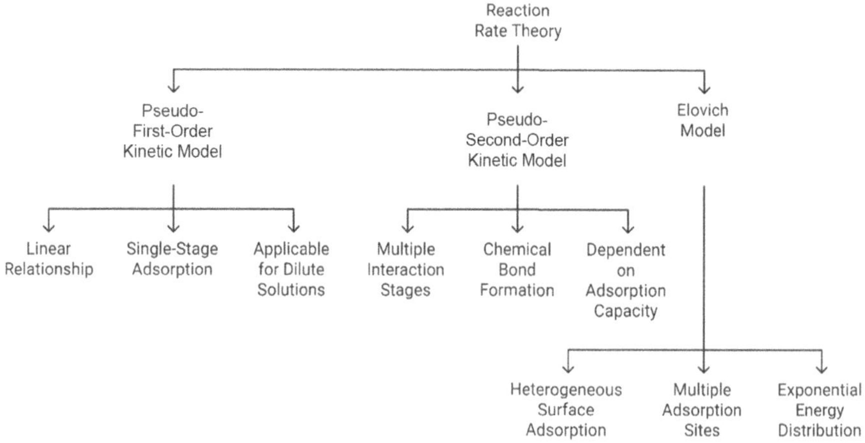

FIGURE 5.11 Comparison of adsorption models in reaction rate theory.

$$\frac{dq_t}{dt} = k_1\left(q_e - q_t\right) \tag{5.5}$$

where q_t is the amount of adsorbate adsorbed at time t, q_e is the adsorption capacity at equilibrium, and k_1 is the rate constant. Integrating Equation 5.5 yields Equation 5.6:

$$\ln\left(q_e - q_t\right) = \ln q_e - k_1 t \tag{5.6}$$

This approach works well in settings where physical interactions are the main mechanism of adsorption. Zhang et al. The adsorption of organic contaminants on activated carbon under low-concentration circumstances was well represented by pseudo-first-order kinetics (Belhachemi, 2021).

5.4.2.2 Pseudo-Second-Order Kinetic Model

The pseudo-second-order model postulates that chemisorption drives adsorption, with the rate being determined by the square of the number of active sites that are not occupied. It is given by Equation 5.7:

$$\frac{dq_t}{dt} = k_2\left(q_e - q_t\right)^2 \tag{5.7}$$

Integrating Equation 5.7 leads to Equation 5.8:

$$\frac{t}{q_t} = \frac{1}{k_2 q_e^2} + \frac{t}{q_e} \tag{5.8}$$

Here, k_2 is the rate constant. Adsorption systems requiring covalent bonding or ion exchange frequently employ this concept. For instance, Zheng et al. (2021) showed that the adsorption of heavy metals on polymer-modified bioadsorbents was well described by pseudo-second-order kinetics (Zheng et al., 2021).

5.4.2.3 Elovich Model

For heterogeneous surfaces, where the adsorption rate drops exponentially with increasing surface coverage, the Elovich model works very well. The model is expressed in Equation 5.9:

$$\frac{dq_t}{dt} = \alpha e^{-\beta q_t} \tag{5.9}$$

Integrating and rearranging:

$$q_t = \frac{1}{\beta}\ln\left(\alpha\beta\right) + \frac{1}{\beta}\text{lnt} \tag{5.10}$$

Here, α represents the initial adsorption rate, and β describes the surface coverage effect. Elovich kinetics have been successfully applied in systems where chemisorption occurs on porous adsorbents (Aharoni and Tompkins, 1970).

5.4.3 Factors Influencing Reaction Rates

The speed of a chemical reaction is influenced by several key factors. Higher concentrations of reactants increase collision frequency, accelerating the reaction (Figure 5.12). Temperature also plays a critical role, as elevated temperatures provide molecules with more energy to overcome activation barriers.

5.4.3.1 Temperature

Adsorption and diffusion rates are increased by higher temperatures because they raise the kinetic energy of the adsorbate molecules. Excessive heating, however, may lessen the adsorption driving power in exothermic adsorption processes. According to a study by Chaudhry et al. (2017), temperature had a major impact on how arsenic adhered to iron-zirconium binary oxide-coated sand (Chaudhry et al., 2017).

5.4.3.2 pH

The charge of adsorbents and adsorbates is influenced by the pH of the solution, which changes how they interact. For instance, metal ion adsorption on negatively

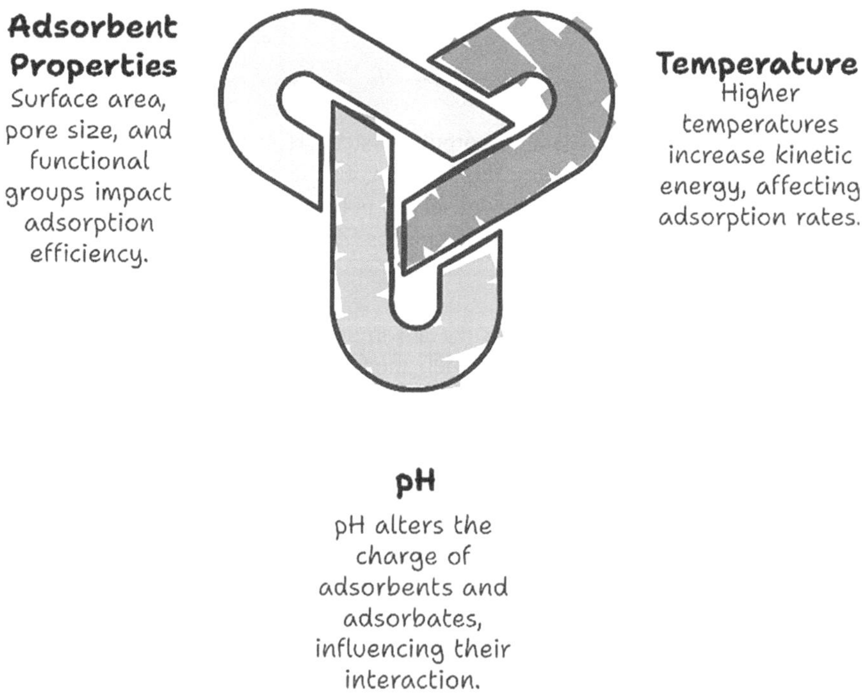

FIGURE 5.12 Factors affecting adsorption in water treatment.

charged surfaces is frequently favored in acidic environments. According to Xu et al. (2022), electrostatic forces caused cadmium to adsorb on clay minerals most effectively at different pH levels (Xu et al., 2022).

5.4.3.3 Adsorbent Properties

Adsorption rates are strongly influenced by surface area, pore size dispersion, and functional groups. Adsorption kinetics are accelerated by functionalized materials with customized active sites or increased porosity. Hussain et al. (2025) pointed out that MOFs and modified biochars are two examples of sophisticated adsorbents with excellent adsorption efficiencies (Hussain et al., 2025).

5.4.4 Applications for Water Treatment

Water treatment system design has benefited greatly from the application of reaction rate theory. Adsorption columns are designed to remove as much pollution as possible using kinetic models. Using catalytic surfaces to increase reaction rates and hasten the breakdown of contaminants is known as catalytic adsorption. To create synergistic effects, hybrid systems integrate adsorption with other processes such as filtration or advanced oxidation. For example, eliminating pesticides from water by combining adsorption and catalytic degradation (Saleh et al., 2020).

From initial diffusion to surface contacts, adsorption kinetics can be fully understood using reaction rate theory. Pseudo-first-order, pseudo-second-order, and Elovich kinetic models are helpful for precisely describing and optimizing adsorption processes. Reaction rate theory continues to be a fundamental component of research and practice in adsorption-based water treatment, especially when combined with new materials and a better comprehension of environmental influences.

5.5 ADSORBENTS USED IN WATER TREATMENT

One of the biggest problems facing the world today is water contamination, which calls for creative and long-lasting cleansing methods. Natural adsorbents have shown great promise in meeting this environmental requirement by providing economical and sustainable substitutes for synthetic treatment techniques. At the complex nexus of material science and environmental engineering, these extraordinary materials efficiently absorb and eliminate contaminants from aqueous systems by utilizing their inherent structural and chemical characteristics.

5.5.1 Natural Adsorbent

Natural adsorbents are eco-friendly materials derived from natural sources like clay, activated carbon, or biomass (Figure 5.13). They effectively capture pollutants due to their porous structures and high surface area. These sustainable options are widely used in water purification and air filtration systems.

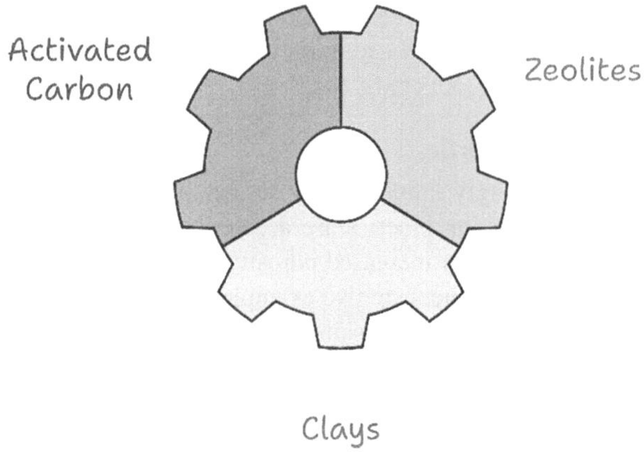

FIGURE 5.13 Natural adsorbents.

5.5.1.1 Activated Carbon

Activated carbon is a natural adsorbent of the highest efficiency. It is characterized by its exceptional surface area and adsorption powers. A rigorous activation process involving thermal or chemical treatments significantly improves the porosity and surface functioning of activated carbon. It is made from carbonaceous materials such as wood, coconut shells, and agricultural waste. The resultant material has a labyrinthine structure with an incredible internal surface area that can surpass 3000 m^2/g, due to an intricate network of micropores and mesopores (Xiaohong Wang et al., 2022). Activated carbon can adsorb a wide variety of pollutants, such as organic compounds, heavy metals, and different ionic species, according to its adsorption mechanism, which is mainly controlled by van der Waals forces and π–π interactions.

5.5.1.2 Zeolite

The distinctive crystalline aluminosilicate structures of zeolites have complex three-dimensional frameworks. It makes them an intriguing class of natural adsorbents. These microporous minerals have exceptional ion-exchange and molecular sieving properties and are found naturally in volcanic areas. They can also be chemically synthesized. Due to their unique porous structure, they enhance selective adsorption of specific molecules based on size and chemical composition, making them highly effective for water treatment.

Zeolites use a combination of chemical and physical interactions in their adsorption mechanism. The cation exchange mechanism is essential for the removal of contaminants. Abdelwahab and Thabet's (2023) research demonstrated the remarkable ability of zeolites to remove radioactive elements, heavy metals, and ammonium ions from aqueous environments. Natural zeolites, such as clinoptilolite, have shown a strong affinity for ions like cesium (Cs^+), ammonium (NH_4^+), and other heavy metals. Under ideal circumstances, their adsorption efficiencies frequently surpass 80% (Abdelwahab and Thabet, 2023).

5.5.1.3 Clays

Clay minerals are an advanced class of naturally occurring adsorbents. They are distinguished by their remarkable surface chemistry and layered phyllosilicate structures. They are made up of substances including bentonite, kaolinite, and montmorillonite. Their distinct structural arrangements allow for exceptional adsorption properties. Their layered structure, which is made up of octahedral and tetrahedral sheets, produces large surface areas and charge fluctuations that enable intricate interactions with water pollutants.

Several processes, such as ion exchange, surface complexation, and electrostatic interactions, are involved in the adsorption mechanism of clays. Bhattacharyya and Gupta's (2008) study showed how clay minerals have a remarkable ability to remove organic pollutants, heavy metals, and developing contaminants from water systems. For example, montmorillonite has demonstrated a remarkable ability to adsorb cationic species such as lead, cadmium, and copper, with adsorption efficiencies varying from 70 to 90% based on environmental factors (Bhattacharyya and Gupta, 2008). These natural adsorbents have the capacity to adsorb pollutants, not only because of their unique qualities but also because of their potential for modification and combinatorial methods. To increase their application and efficiency in water treatment technologies, researchers are looking more closely at chemically modified natural adsorbents and hybrid systems.

5.5.2 Synthetic Adsorbent

Synthetic adsorbents are now at the forefront of environmental remediation techniques due to the unrelenting pursuit of increasingly sophisticated water purification technology. Synthetic adsorbents can be carefully crafted to tackle complex water contamination problems, in contrast to their natural counterparts. By combining chemical design, material science, and environmental sustainability, these advanced synthetic materials address certain contaminants more effectively than any other adsorbent. There are various synthetic adsorbents listed in Figure 5.14.

5.5.3 Synthetic Resins

Synthetic resins are distinguished by their remarkable functional group variety and complex polymer networks, making them advanced molecular warriors in the field of water treatment. These tailored materials enable unmatched selectivity and customization for targeted pollutant removal. Specifically, ion-exchange resins are exceptional synthetic adsorbents with functional groups that are precisely engineered to allow for highly selective interactions with contaminating molecules.

Unlike conventional adsorbents, synthetic resins use complicated chemical interactions in their adsorption mechanism that go beyond traditional adsorbents. Heavy metals, radioactive ions, and developing pharmaceutical residues are among the ionic species for which ion-exchange resins can achieve contaminant removal rates surpassing 95%, according to research by Shamshad and Ur Rehman (2024). Researchers may now develop resins with previously unheard-of specificity for difficult contaminants like rare earth elements and complex organic molecules because of their modular design, which enables strategic functionalization.

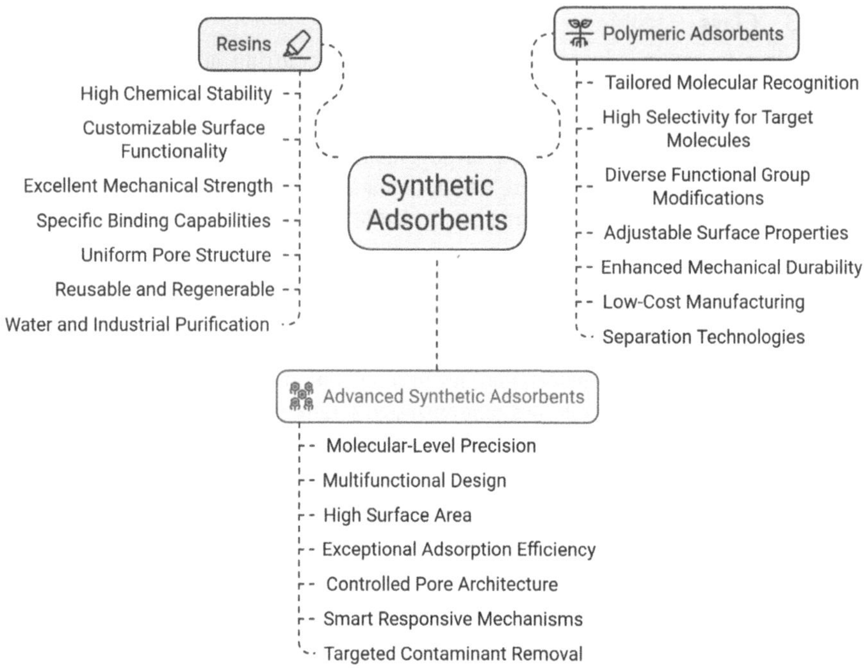

FIGURE 5.14 Synthetic adsorbent.

5.5.4 Polymeric Adsorbents

Polymeric adsorbents also adsorb efficiently because of their extremely complex molecular structures. The structural complexity of these materials is carefully considered throughout their creation, and they have complex three-dimensional networks that optimize surface area and interaction potential with the chosen pollutants. Unlike conventional adsorbents, polymeric materials can be molecularly designed to provide hitherto unheard-of levels of customization in their adsorption properties.

Polymeric adsorbents have a special adsorption mechanism that combines a variety of physical and chemical interactions, such as electrostatic attractions, hydrogen bonds, and π–π stacking. The extraordinary capacity of polymeric adsorbents to remove developing pollutants such as microplastics and pharmaceutical residues was proven in a landmark work by Hazarika et al. (2023). Under a variety of environmental circumstances, the removal efficiencies ranged from 85 to 98% (Hazarika et al., 2023). Their adaptability extends beyond traditional water treatment paradigms, offering potential solutions for increasingly sophisticated contamination challenges.

5.5.5 Advanced Synthetic Adsorbents

Researchers are investigating more and more novel materials that challenge the limits of conventional water treatment technologies. Magnetic nanocomposite adsorbents are a novel method that combines remarkable adsorption properties with magnetic responsiveness. These materials enable recovery and reusability of adsorbents, since they are easily separated from treated water using magnetic fields.

With incredibly intricate internal structures that optimize surface interactions, MOFs and hierarchical porous materials represent the next generation of synthetic adsorbents. MOFs have the potential to capture difficult contaminants such as organic pollutants and heavy metals, according to research by Bedia et al. (2020). Some compounds have removal efficiencies that are almost at the level of total eradication. It is a paradigm leap in water treatment technology. Synthetic adsorbents can form an integrated and flexible component for water treatment systems. Multifunctional adsorbents, which can simultaneously address several pollutant classes, are being investigated by researchers more and more as a comprehensive strategy to water purification that goes beyond conventional treatment techniques.

5.6 EMERGING ADSORBENTS

The field of water treatment technology is undergoing a significant transition as new adsorbents redefine how we treat environmental contamination. These innovative materials bridge the gap between scientific innovation and environmental sustainability, enabling us to tackle difficult water contamination concerns. New adsorbents are emerging as molecular protectors for our most valuable water resource.

5.6.1 Nanomaterials

Nanomaterials provide a remarkable method of treating water that works at the most basic level. Usually measuring between 1 and 100 nanometers, these microscopic wonders have special physicochemical characteristics that set them apart from traditional adsorbents. By permitting interactions that were previously unthinkable in conventional water treatment techniques, their unparalleled surface area-to-volume ratio establishes a ground-breaking platform for pollutant capture.

By utilizing quantum-scale interactions, the adsorption mechanism of nanomaterials surpasses traditional material limits. Advanced nanomaterials could remove complex contaminants such as heavy metals, microplastics, and developing pharmaceutical residues with contaminant removal efficiency of above 99%, according to research by Hazarika et al. (2023). The ability of carbon-based nanomaterials, such as graphene oxide and carbon nanotubes, to capture a wide variety of contaminants through a variety of interaction mechanisms, such as surface complexation, π–π interactions, and electrostatic attraction, has particularly captivated researchers.

5.6.2 Bioadsorbents

Bioadsorbents are a revolutionary method of treating water that uses the inherent properties of biological materials to solve environmental problems. Originating from marine biomass, microbiological sources, and agricultural waste, these novel materials offer a sustainable substitute for synthetic adsorbents. Through millions of years of natural selection, their distinct molecular architectures have developed to provide complex mechanisms for pollutant collection that are on par with the most cutting-edge synthetic technologies.

Ion exchange, surface complexation, molecule recognition, and other intricate biological and chemical interactions are all part of the biosorbent's adsorption mechanism. Vijayaraghavan and Yun (2008) conducted ground-breaking work that demonstrated the remarkable potential of biomass-derived adsorbents, with removal efficiencies ranging from 85 to 97% for a variety of organic contaminants and heavy metals (Vijayaraghavan and Yun, 2008). Materials such as carbons generated from agricultural waste, alginate from marine algae, and chitosan from crab shells are examples of the extraordinary diversity and potential of bioadsorbents.

5.6.3 Hybrid and Multifunctional Emerging Adsorbents

The most intriguing advancements in newly developed adsorbents are found in their capacity for multifunctionality and hybridization. To create complex molecular systems that can handle challenging contamination issues, researchers are increasingly investigating composite materials that combine the advantages of several adsorbent classes. For example, magnetic nanocomposite bioadsorbents enable previously unexplored water treatment capabilities by combining the sustainability of biological materials with the accuracy of nanoparticles.

Covalent organic frameworks (COFs) and MOFs are two more cutting-edge adsorbent technology developments. These materials offer selectivity and utility due to their extraordinarily intricate three-dimensional architectures, which can be precisely designed at the molecular level. MOFs could capture numerous contaminant classes at the same time (Hu et al., 2022).

Emerging adsorbents have more potential than classical adsorbents in water purification. By providing flexible, sustainable solutions that can be customized to specific pollution scenarios. These materials mark a paradigm shift in the way to tackle environmental concerns. New adsorbents are poised to transform our knowledge of water filtration as research continues to push the limits of molecular engineering.

5.7 APPLICATIONS IN WATER TREATMENT

Water contamination is one of the most important environmental issues in our world. It requires complex and creative removal techniques that take into account a variety of pollutant classes. The complex field of water treatment technologies has seen a significant shift from straightforward filtering methods to extremely complex molecular-level interventions. To address important water quality issues, this chapter examines the various uses of cutting-edge adsorption technologies, revealing the intricate processes that support successful environmental restoration.

5.7.1 Removal of Heavy Metals

An overlooked environmental disaster, heavy metal poisoning poses a serious risk to human health and natural systems. Adsorption-based heavy metal removal is a sophisticated molecular intervention technique that uses complex chemical interactions to adsorb and render these persistent contaminants immobile. The crucial significance of focused removal tactics was emphasized by research by Feisal et al. (2024), which showed that, in the right circumstances, sophisticated adsorbents may remove heavy metals with efficiencies of over 95%.

Surface complexation, ion exchange, and electrostatic attraction are some of the intricate surface interactions that are a part of the heavy metal adsorption mechanism. Lead, cadmium, mercury, and chromium can be selectively captured by materials including MOFs, functionalized nanomaterials, and modified activated carbons. Innovative approaches like magnetic nanocomposite adsorbents enable not just capture but also efficient recovery, representing a paradigm shift in heavy metal remediation technologies.

5.7.2 Organic Pollutant Removal

Organic contaminants, with their varied chemical compositions and enduring effects on the ecosystem, pose a significant barrier to water treatment. The advanced adsorption method traps complex organic compounds, and has become a crucial intervention technique. The elimination mechanism uses advanced molecular identification and interaction techniques.

Innovative research by Belhachemi (2021) showed that functionalized polymeric adsorbents and improved carbon-based materials may achieve 92–98% removal efficiency for organic pollutants. Particularly difficult contaminant classifications that require creative removal techniques are pesticides, pharmaceutical residues, and industrial chemical intermediates. New technologies that provide tailored solutions, such as graphene-based composites and molecularly imprinted polymers, allow for the remarkably precise selective capture of organic molecules.

5.7.3 Removal of Dyes and Pigments

The manufacturing and textile sectors produce large amounts of chromatic pollution. They introduce intricate dye molecules that present serious environmental problems. Since they can capture and neutralize these persistent colorants with previously unknown capabilities, advanced adsorption technologies have become a sophisticated intervention technique. The removal method employs techniques that go beyond straightforward physical filtering and entail complex interactions between dye molecules and adsorbent surfaces.

Modified activated carbons, agricultural waste products, and sophisticated nanomaterials have demonstrated exceptional promise in tackling this intricate environmental issue by providing long-term and economical intervention techniques.

5.7.4 Comprehensive Case Studies

Advanced adsorption technologies have remarkable potential in solving water treatment problems. In a seminal case study, the effectiveness of combined adsorption methods in comprehensive water cleanup was shown in a Southeast Asian industrial manufacturing sector. Heavy metals, organic pollutants, and chromatic impurities were almost entirely removed by researchers using a multi-stage treatment technique with nanomaterial-enhanced adsorbents. Advanced polymeric adsorbents were used to handle emerging pollutants such as microplastics and pharmaceutical residues at a European city water treatment facility, providing another compelling example. With pollutant removal efficiency above 95%, the integrated treatment system marked a breakthrough in urban water purification techniques. These practical applications highlight how advanced adsorption technologies can have a profound impact (Feisal et al., 2024).

5.7.5 Emerging Frontiers and Future Perspectives

The development of sophisticated and adaptive adsorption systems that can react dynamically to intricate contamination scenarios is crucial for the future of water treatment technology. In order to create materials with molecular precision and versatility, researchers are investigating artificial intelligence-driven adsorbent design. Innovative methods of purifying water are promised by the fusion of environmental engineering, materials science, and nanotechnology.

Adsorbent development is increasingly guided by sustainable and circular design concepts, with researchers concentrating on materials made from recyclable synthetic compounds, renewable biomass, and agricultural waste. All this encompassing strategy takes into account the wider environmental impact of water treatment technologies in addition to the removal of contaminants (Shamshad and Ur Rehman, 2024).

5.8 CHALLENGES AND FUTURE PERSPECTIVES

Water treatment technology is confronting various difficulties that need many revolutionary solutions. Placing it at a crucial nexus of scientific advancement and environmental necessity. Adsorption technologies must simultaneously push the limits of molecular engineering and sustainable solutions while navigating a maze of scientific, financial, and environmental constraints as the world's water resources grow more susceptible to complicated contamination.

5.8.1 Technological and Methodological Challenges

The contemporary water treatment ecosystem confronts a multifaceted array of challenges that extend far beyond traditional contaminant removal strategies. Emerging pollutants, including pharmaceutical residues, microplastics, and persistent organic compounds, represent a dynamic and evolving threat that conventional adsorption technologies struggle to comprehensively address. Research by Rodriguez-Narvaez et al. (2017) highlighted the complexity of these emerging contaminants, demonstrating that traditional adsorption mechanisms often fall short in capturing the intricate molecular structures of modern pollutants. The scalability of advanced adsorption technologies presents another significant hurdle. While laboratory-scale experiments demonstrate remarkable removal efficiencies, translating these innovations into large-scale, economically viable water treatment solutions remains a formidable challenge. A significant obstacle to the broad adoption of advanced adsorbent technologies is the financial limitations associated with their implementation, especially when those technologies involve nanomaterials and complex molecular engineering. Cost-effective production techniques and long-term performance evaluation of improved adsorbent materials are crucial, according to Satyam and Patra (2024).

5.8.2 Environmental and Sustainability Considerations

As adsorption technologies advance, sustainability becomes a critical consideration. Manufacturing, use, and disposal of adsorbent material have an impact on the environment that needs to be carefully considered. Increasingly, researchers are concentrating on creating closed-loop systems that reduce waste production and maximize material reusability. Advanced water treatment technologies are increasingly incorporating the idea of the circular economy, which challenges conventional linear approaches to material consumption.

Another crucial area of research is the regeneration and long-term performance of adsorbent materials. After several cycles of usage, many sophisticated adsorbents exhibit notable performance degradation, which restricts their usefulness in actual water treatment situations. Bagotia (2025) investigated novel regeneration strategies that might increase the useful life of sophisticated adsorbent materials and provide encouraging new information for more environmentally friendly water treatment methods.

5.8.3 Interdisciplinary Integration and Technological Convergence

Adsorption technology for water treatment depends on technological convergence and interdisciplinary cooperation. According to recent studies, the best answers will come from combining several scientific fields, such as molecular biology, environmental engineering, materials science, and nanotechnology.

To improve adsorbent design and performance prediction, artificial intelligence and machine learning are being used more and more. The development of next-generation adsorbent technologies is currently accelerated by sophisticated computer models that can accurately anticipate adsorption characteristics and simulate molecular interactions. Dansawad et al.'s (2023) study showed how AI-driven methods might be used to create targeted adsorbent materials with previously unheard-of specificity.

5.8.4 Future Perspectives and Transformative Potential

Adsorption technologies for water treatment have a bright future, thanks to their remarkable potential for game-changing innovation. Researchers are looking into more advanced methods that go beyond the conventional removal of contaminants, such as smart adsorbent systems that can selectively capture molecules and recognize them in real time. The amazing chemical interactions seen in natural systems can serve as the basis for bioinspired design methodologies, which have the potential to revolutionize water purification technology.

A further potential breakthrough area is the combination of adaptable adsorbent materials with sensor technologies and sophisticated monitoring systems. Consider adsorbent systems that, in response to particular pollution profiles, can dynamically modify their molecular configuration. This concept shifts from static filtering techniques to an ecosystem for intelligent, adaptive water treatment.

5.9 CONCLUSION

Water treatment technologies are at a pivotal juncture when environmental demands from around the world meet scientific advancement. A complicated story of chemical interactions, technological advancements, and the unwavering human quest for environmental sustainability is revealed by exploring adsorption mechanisms and kinetics. Our investigation has spanned the complicated terrain of adsorbent technologies, revealing the deep intricacy behind what seems to be a straightforward pollutant removal procedure.

According to research by Renu and Sithole (2024), the development of water treatment technology is essentially a story of innovation and adaptation. More than just a technical development, the move from conventional filtering techniques to complex molecular identification systems marks a fundamental shift in how we address environmental issues. Whether carefully crafted in cutting-edge labs or sourced from natural sources, every adsorbent material has the capacity to handle progressively more complicated contamination situations.

The enormous intricacy of molecular interactions is revealed by the flexibility of adsorption mechanisms. While synthetic and new adsorbents push the limits of what was previously thought to be feasible, natural adsorbents such as zeolites and activated carbon show off nature's innate sophistication. The intricate interplay of electrostatic attractions, π–π interactions, and surface complexation mechanisms illustrates the nuanced world of molecular capture, where seemingly simple interactions yield remarkable purification capabilities.

At the forefront of this technological revolution are nanomaterials and bioadsorbents, which provide a higher level of precision in the removal of contaminants. The remarkable potential of these sophisticated materials was demonstrated in a groundbreaking work by Feisal et al. (2024), which showed removal efficiencies that resulted in almost complete eradication of complex contaminants. Water treatment is seen as a complex molecular conversation rather than a mechanical filtration procedure, and these materials reflect a new philosophical approach to environmental rehabilitation that goes beyond simple technological fixes. Researchers are able to predict and manipulate molecular interactions with surprising accuracy. This approach transforms water treatment from a reactive process to a proactive, intelligent system capable of anticipating and addressing contamination challenges.

The road ahead is not without major obstacles. The complicated environmental impact of new materials, scalability issues, and economic constraints necessitate ongoing innovation and critical analysis. The need to create economical, sustainable strategies that strike a balance between technological sophistication and real-world application is highlighted by the work of Gadore and Ahmaruzzaman (2021). The ultimate objective goes beyond technological advancement and includes developing comprehensive and ecologically conscious solutions that may be applied everywhere.

These developments have far-reaching effects that go well beyond water purification. Potential uses for the molecular knowledge and technological advancements made in this area include resource management, medicinal technology, and environmental cleanup. Every development in adsorption processes is a step toward more intelligent and sustainable interactions with our surroundings, as well as a scientific triumph.

At this pivotal moment, the future of water treatment seems both extremely bright and full of challenges. The integration of environmental science, computational technology, and molecular engineering presents possibilities to tackle the world's water quality issues. At its core, the adsorption story is a tale of human potential.

REFERENCES

Abdelwahab, O. and Thabet, W. M., Natural Zeolites and Zeolite Composites for Heavy Metal Removal from Contaminated Water and Their Applications in Aquaculture Systems: A Review, *Egyptian Journal of Aquatic Research*, vol. 49, no. 4, pp. 431–443, 2023. DOI: 10.1016/J.EJAR.2023.11.004

Aharoni, C. and Tompkins, F. C., Kinetics of Adsorption and Desorption and the Elovich Equation, *Advances in Catalysis*, vol. 21, no. C, pp. 1–49, 1970. DOI: 10.1016/S0360-0564(08)60563-5

Ahmad, M., Rajapaksha, A. U., Lim, J. E., Zhang, M., Bolan, N., Mohan, D., Vithanage, M., Lee, S. S. and Ok, Y. S., Biochar as a Sorbent for Contaminant Management in Soil and Water: A Review, *Chemosphere*, vol. 99, pp. 19–33, 2014. DOI: 10.1016/J.CHEMOSPHERE.2013.10.071

Alafnan, S., Awotunde, A., Glatz, G., Adjei, S., Alrumaih, I. and Gowida, A., Langmuir Adsorption Isotherm in Unconventional Resources: Applicability and Limitations, *Journal of Petroleum Science and Engineering*, vol. 207, p. 109172, 2021. DOI: 10.1016/J.PETROL.2021.109172

Al-Amrani, W. A. and Onaizi, S. A., Adsorptive Removal of Heavy Metals from Wastewater Using Emerging Nanostructured Materials: A State-of-the-Art Review, *Separation and Purification Technology*, vol. 343, p. 127018, 2024. DOI: 10.1016/J.SEPPUR.2024.127018

AlAqad, K. M., Abdelnaby, M. M., Tanimu, A., Abdulazeez, I. and Elsharif, A. M., Adsorbent Materials for Water Treatment: A Review of Current Trends and Future Challenges, *Environmental Pollution and Management*, vol. 2, pp. 1–13, 2025. https://linkinghub.elsevier.com/retrieve/pii/S2950305124000330. DOI: 10.1016/J.EPM.2024.12.003

Al-Ghouti, M. A. and Da'ana, D. A., Guidelines for the Use and Interpretation of Adsorption Isotherm Models: A Review, *Journal of Hazardous Materials*, vol. 393, p. 122383, 2020. DOI: 10.1016/J.JHAZMAT.2020.122383

Badsha, M. A. H., Khan, M., Wu, B., Kumar, A. and Lo, I. M. C., Role of Surface Functional Groups of Hydrogels in Metal Adsorption: From Performance to Mechanism, *Journal of Hazardous Materials*, vol. 408, p. 124463, 2021. DOI: 10.1016/J.JHAZMAT.2020.124463

Bagotia, N., Regeneration Strategies for Exhausted Adsorbents Used in Water Treatment - A Critical Review, *Journal of Water, Process Engineering*, vol. 69, p. 106560, 2025. DOI: 10.1016/J.JWPE.2024.106560

Bedia, J., Muelas-Ramos, V., Peñas-Garzón, M., Gómez-Avilés, A., Rodriguez, J. J. and Belver, C., Metal–Organic Frameworks for Water Purification, *Nanomaterials for the Detection and Removal of Wastewater Pollutants*, pp. 241–283, 2020. DOI: 10.1016/B978-0-12-818489-9.00009-8

Belhachemi, M., Adsorption of Organic Compounds on Activated Carbons, *Sorbents Materials for Controlling Environmental Pollution: Current State and Trends*, pp. 355–385, 2021. DOI: 10.1016/B978-0-12-820042-1.00006-7

Bhattacharyya, K. G. and Gupta, S. S., Adsorption of a Few Heavy Metals on Natural and Modified Kaolinite and Montmorillonite: A Review, *Advances in Colloid and Interface Science*, vol. 140, no. 2, pp. 114–131, 2008. DOI: 10.1016/J.CIS.2007.12.008

Blesa, M. A., Weisz, A. D., Morando, P. J., Salfity, J. A., Magaz, G. E. and Regazzoni, A. E., The Interaction of Metal Oxide Surfaces With Complexing Agents Dissolved in Water, *Coordination Chemistry Reviews*, vol. 196, no. 1, pp. 31–63, 2000. DOI: 10.1016/S0010-8545(99)00005-3

Chaudhry, S. A., Zaidi, Z. and Siddiqui, S. I., Isotherm, Kinetic and Thermodynamics of Arsenic Adsorption onto Iron-Zirconium Binary Oxide-Coated Sand (IZBOCS): Modelling and Process Optimization, *Journal of Molecular Liquids*, vol. 229, pp. 230–240, 2017. DOI: 10.1016/J.MOLLIQ.2016.12.048

Chmelik, C. and Kärger, J., Molecular Transport in Nanoporous Materials, *Nanoporous Materials for Molecule Separation and Conversion*, pp. 169–215, 2020. DOI: 10.1016/B978-0-12-818487-5.00006-6

Colozza, N., Mattiello, A., Duranti, L., Cerra, S., Cartoni, A. and Arduini, F., Removal of Multiple Metals From Real Wastewater Combining Sludges With Carbon Black and Chitosan: Integrating Sustainable Remediation and Waste Recycling, *Journal of Environmental Chemical Engineering*, p. 116660, 2025. https://linkinghub.elsevier.com/retrieve/pii/S2213343725013569. DOI: 10.1016/J.JECE.2025.116660

Dansawad, P., Li, Y., Li, Y., Zhang, J., You, S., Li, W. and Yi, S., Machine Learning Toward Improving the Performance of Membrane-Based Wastewater Treatment: A Review, *Advanced Membranes*, vol. 3, p. 100072, 2023. DOI: 10.1016/J.ADVMEM.2023.100072

Dudgeon, D., Arthington, A. H., Gessner, M. O., Kawabata, Z. I., Knowler, D. J., Lévêque, C., Naiman, R. J., et al., Freshwater Biodiversity: Importance, Threats, Status and Conservation Challenges, *Biological Reviews of the Cambridge Philosophical Society*, 2006, vol. 81, no. 2, 163–182.

Falkenmark, M., and Rockström, J., The New Blue and Green Water Paradigm: Breaking New Ground for Water Resources Planning and Management Water for Food and Hunger Alleviation, *Journal of Water Resources Planning and Management*, vol. 132, n.d. DOI: 10.1061/(ASCE)0733-9496(2006)132:3(129)

Feisal, N. A. S., Kamaludin, N. H., Ahmad, M. A. and Tengku Ibrahim, T. N. B., A Comprehensive Review of Nanomaterials for Efficient Heavy Metal Ions Removal in Water Treatment, *Journal of Water Process Engineering*, vol. 64, p. 105566, 2024. DOI: 10.1016/J.JWPE.2024.105566

Foo, K. Y. and Hameed, B. H., Insights into the Modeling of Adsorption Isotherm Systems, *Chemical Engineering Journal*, vol. 156, no. 1, pp. 2–10, 2010. DOI: 10.1016/J.CEJ.2009.09.013

Forstinus, N., Ikechukwu, N., Emenike, M. and Christiana, A., Water and Waterborne Diseases: A Review, *International Journal of Tropical Disease & Health*, vol. 12, no. 4, pp. 1–14, 2016. DOI: 10.9734/ijtdh/2016/21895

Gadore, V. and Ahmaruzzaman, M., Smart Materials for Remediation of Aqueous Environmental Contaminants, *Journal of Environmental Chemical Engineering*, vol. 9, no. 6, p. 106486, 2021. DOI: 10.1016/J.JECE.2021.106486

Gawande, S. M., Belwalkar, N. S. and Mane, A. A., Adsorption and Its Isotherm – Theory, *International Journal of Engine Research*, vol. 6, no. 6, p. 312, 2017. DOI: 10.5958/2319-6890.2017.00026.5

Gilormini, P. and Verdu, J., On the Role of Hydrogen Bonding on Water Absorption in Polymers, *Polymer*, vol. 142, pp. 164–169, 2018. DOI: 10.1016/J.POLYMER.2018.03.033

Goh, S. W., Ng, Q. H., Rahim, S. K. E. A., Low, S. C., Hoo, P. Y., Yeo, R. Y. Z., Chew, T. L. and Jawad, Z. A., Recent Advancements in Smart Materials for the Removal of Organic, Inorganic and Microbial Pollutants in Water Treatment: A Review, *Journal of Water Process Engineering*, vol. 70, p. 106993, 2025. DOI: 10.1016/J.JWPE.2025.106993

Gupta, V. K. and Suhas, Application of Low-Cost Adsorbents for Dye Removal – A Review, *Journal of Environmental Management*, vol. 90, no. 8, pp. 2313–2342, 2009. DOI: 10.1016/J.JENVMAN.2008.11.017

Hattori, H., Solid Base Catalysts: Fundamentals and Their Applications in Organic Reactions, *Applied Catalysis A: General*, vol. 504, pp. 103–109, 2015. DOI: 10.1016/J.APCATA.2014.10.060

Hazarika, B., Ahmaruzzaman, M., Santosh, M. S., Barceló, D. and Rtimi, S., Advances in Polymer-Based Nanocomposite Membranes for Water Remediation: Preparation Methods, Critical Issues and Mechanisms, *Journal of Environmental Chemical Engineering*, vol. 11, no. 6, p. 111401, 2023. DOI: 10.1016/J.JECE.2023.111401

Ho, Y. S. and McKay, G., A Comparison of Chemisorption Kinetic Models Applied to Pollutant Removal on Various Sorbents, *Process Safety and Environmental Protection*, vol. 76, no. 4, pp. 332–340, 1998. DOI: 10.1205/095758298529696

Hong, S., Vincenzo, M. Di, Tiraferri, A., Bertozzi, E., Górecki, R., Davaasuren, B., Li, X. and Nunes, S. P., Precision Ion Separation via Self-Assembled Channels, *Nature Communications*, vol. 15, no. 1, 2024. DOI: 10.1038/s41467-024-47083-0

Hu, T., Tang, L., Feng, H., Zhang, J., Li, X., Zuo, Y., Lu, Z. and Tang, W., Metal-Organic Frameworks (MOFs) and Their Derivatives as Emerging Catalysts for Electro-Fenton Process in Water Purification, *Coordination Chemistry Reviews*, vol. 451, p. 214277, 2022. DOI: 10.1016/J.CCR.2021.214277

Huang, C. H., Chang, K. P., Ou, H. De, Chiang, Y. C. and Wang, C. F., Adsorption of Cationic Dyes onto Mesoporous Silica, *Microporous and Mesoporous Materials*, vol. 141, no. 1–3, pp. 102–109, 2011. DOI: 10.1016/J.MICROMESO.2010.11.002

Hussain, M., Riaz, A., Zeb, H., Ali, A., Mujahid, R., Ahmad, F. and Zafar, M. S., Paving the Path to Water Security: The Role of Advanced Adsorbents in Wastewater Treatment, *Journal of Water Process Engineering*, vol. 71, p. 107333, 2025. DOI: 10.1016/J.JWPE.2025.107333

Kachel, S. R., Klein, B. P., Morbec, J. M., Schöniger, M., Hutter, M., Schmid, M., Kratzer, P., Meyer, B., Tonner, R. and Gottfried, J. M., Chemisorption and Physisorption at the Metal/Organic Interface: Bond Energies of Naphthalene and Azulene on Coinage Metal Surfaces, *Journal of Physical Chemistry C*, vol. 124, no. 15, pp. 8257–8268, 2020. DOI: 10.1021/acs.jpcc.0c00915

Karimi, S., Tavakkoli Yaraki, M. and Karri, R. R., A Comprehensive Review of the Adsorption Mechanisms and Factors Influencing the Adsorption Process From the Perspective of Bioethanol Dehydration, *Renewable and Sustainable Energy Reviews*, vol. 107, pp. 535–553, 2019. DOI: 10.1016/J.RSER.2019.03.025

Khasawneh, O. F. S. and Palaniandy, P., Occurrence and Removal of Pharmaceuticals in Wastewater Treatment Plants, *Process Safety and Environmental Protection*, vol. 150, pp. 532–556, 2021. DOI: 10.1016/J.PSEP.2021.04.045

Kumar, A. and Xagoraraki, I., Pharmaceuticals, Personal Care Products and Endocrine-Disrupting Chemicals in U.S. Surface and Finished Drinking Waters: A Proposed Ranking System, *Science of the Total Environment*, vol. 408, no. 23, pp. 5972–5989, 2010. DOI: 10.1016/J.SCITOTENV.2010.08.048

Lappala, A., The Next Revolution in Computational Simulations: Harnessing AI and Quantum Computing in Molecular Dynamics, *Current Opinion in Structural Biology*, vol. 89, p. 102919, 2024. DOI: 10.1016/J.SBI.2024.102919

Li, M., Xu, Y., Fu, Q., Singh, V. P., Liu, D. and Li, T., Efficient Irrigation Water Allocation and Its Impact on Agricultural Sustainability and Water Scarcity under Uncertainty, *Journal of Hydrology*, vol. 586, p. 124888, 2020. DOI: 10.1016/J.JHYDROL.2020.124888

Li, X., Sun, M., Rooke, J. C., Chen, L. and Su, B. L., Synthesis and Applications of Hierarchically Porous Catalysts, *Chinese Journal of Catalysis*, vol. 34, no. 1, pp. 22–47, 2013. DOI: 10.1016/S1872-2067(11)60507-X

Liu, Y., Some Consideration on the Langmuir Isotherm Equation, *Colloids and Surfaces A: Physicochemical and Engineering Aspects*, vol. 274, no. 1–3, pp. 34–36, 2006. DOI: 10.1016/J.COLSURFA.2005.08.029

Low, M. J. D., Kinetics of Chemisorption of Gases on Solids, *Chemical Reviews*, vol. 60, no. 3, pp. 267–312, 1960.

Lu, G., Wang, G., Wang, P. H., Yang, Z., Yan, H., Ni, W., Zhang, L. and Yan, Y. M., Enhanced Capacitive Deionization Performance With Carbon Electrodes Prepared With a Modified Evaporation Casting Method, *Desalination*, vol. 386, pp. 32–38, 2016. DOI: 10.1016/J.DESAL.2016.02.008

Mahmood Aljamali, N. and Obaid Alfatlawi, I., Physical and Chemical Adsorption and Its Applications, 2021. http://chemical.journalspub.info/index.php?journal=JTCK&page=index

Mestri, S., Le, T. P. P., Jouini, M. S., AlMarzooqi, F., Hasan, S. W. and Alhseinat, E., Surface Modification of Nanofiltration Membrane for Enhancing Salt Rejection and Scaling Remediation in Brackish Groundwater Desalination Application, *Desalination*, vol. 598, p. 118417, 2025. DOI: 10.1016/J.DESAL.2024.118417

Mizuno, N., & Misono, M., Functionalization of Light Alkanes Catalyzed by Heteropoly Compounds, *Catalytic Activation and Functionalisation of Light Alkanes: Advances and Challenges*, pp. 311–331, 1998. Springer Netherlands, Dordrecht.

Mozaffari Majd, M., Kordzadeh-Kermani, V., Ghalandari, V., Askari, A. and Sillanpää, M., Adsorption Isotherm Models: A Comprehensive and Systematic Review (2010–2020), *Science of the Total Environment*, vol. 812, p. 151334, 2022. DOI: 10.1016/J.SCITOTENV.2021.151334

Park, J., Lee, Y. and Kim, J., Multi-Modal Conditional Diffusion Model Using Signed Distance Functions for Metal-Organic Frameworks Generation, *Nature Communications*, vol. 16, no. 1, 2025. DOI: 10.1038/s41467-024-55390-9

Peng, Z., Li, M., Liu, Y., Yin, J., Fang, H. and Wen, J., Spatiotemporal Evolution and Influencing Factors of Carbon Stock in the Water Receiving Areas from the Perspective of Carbon Neutrality, *Scientific Reports*, vol. 15, no. 1, p. 13003, 2025. https://www.nature.com/articles/s41598-025-96033-3. DOI: 10.1038/s41598-025-96033-3

Pullumbi, P., Brandani, F. and Brandani, S., Gas Separation by Adsorption: Technological Drivers and Opportunities for Improvement, *Current Opinion in Chemical Engineering*, vol. 24, pp. 131–142, 2019. DOI: 10.1016/J.COCHE.2019.04.008

Renu and Sithole, T., A Review on Regeneration of Adsorbent and Recovery of Metals: Adsorbent Disposal and Regeneration Mechanism, *South African Journal of Chemical Engineering*, vol. 50, pp. 39–50, 2024. DOI: 10.1016/J.SAJCE.2024.07.006

Revellame, E. D., Fortela, D. L., Sharp, W., Hernandez, R. and Zappi, M. E., Adsorption Kinetic Modeling Using Pseudo-First Order and Pseudo-Second Order Rate Laws: A Review, *Cleaner Engineering and Technology*, vol. 1, p. 100032, 2020. DOI: 10.1016/J.CLET.2020.100032

Rodriguez-Narvaez, O. M., Peralta-Hernandez, J. M., Goonetilleke, A. and Bandala, E. R., Treatment Technologies for Emerging Contaminants in Water: A Review, *Chemical Engineering Journal*, vol. 323, pp. 361–380, 2017. DOI: 10.1016/J.CEJ.2017.04.106

Saleh, I. A., Zouari, N. and Al-Ghouti, M. A., Removal of Pesticides from Water and Wastewater: Chemical, Physical and Biological Treatment Approaches, *Environmental Technology and Innovation*, vol. 19, p. 101026, 2020. DOI: 10.1016/J.ETI.2020.101026

Satyam, S. and Patra, S., Innovations and Challenges in Adsorption-Based Wastewater Remediation: A Comprehensive Review, *Heliyon*, vol. 10, no. 9, p. e29573, 2024. DOI: 10.1016/J.HELIYON.2024.E29573

Shamshad, J. and Ur Rehman, R., Innovative Approaches to Sustainable Wastewater Treatment: A Comprehensive Exploration of Conventional and Emerging Technologies, *Environmental Science Advances*, vol. 4, no. 2, pp. 189–222, 2024. DOI: 10.1039/D4VA00136B

Shao, X., Bian, Z., Li, B., Zhan, F., Cheng, X., Shen, Y., Li, Z., Zhou, Q., Cai, R. and Feng, C., Enhanced Mass Transport on Single-Atom Ni-N-C Catalysts with Hierarchical Pore Structures for Efficient CO2 Electroreduction, *Separation and Purification Technology*, vol. 359, p. 130576, 2025. DOI: 10.1016/J.SEPPUR.2024.130576

Sharma, A., Kumar, V., Shahzad, B., Ramakrishnan, M., Singh Sidhu, G. P., Bali, A. S., Handa, N., et al., Photosynthetic Response of Plants Under Different Abiotic Stresses: A Review, *Journal of Plant Growth Regulation*, vol. 39, no. 2, pp. 509–531, 2020.

Shimizu, S. and Matubayasi, N., Cooperative Sorption on Porous Materials, *Langmuir*, vol. 37, no. 34, pp. 10279–10290, 2021. DOI: 10.1021/acs.langmuir.1c01236

Sing, K. S. W., The Use of Physisorption for the Characterization of Microporous Carbons, *Carbon*, vol. 27, no. 1, pp. 5–11, 1989. DOI: 10.1016/0008-6223(89)90151-6

Skopp, J., Derivation of the Freundlich Adsorption Isotherm from Kinetics, *Journal of Chemical Education*, vol. 86, no. 11, p. 1341, 2009.

Snyder, L. R., The Role of Dispersion Interactions in Adsorption Chromatography, *Journal of Chromatography A*, vol. 36, no. C, pp. 455–475, 1968. DOI: 10.1016/S0021-9673(01)92975-9

Swenson, H. and Stadie, N. P., Langmuir's Theory of Adsorption: A Centennial Review, *Langmuir*, vol. 35, no. 16, pp. 5409–5426, 2019. DOI: 10.1021/acs.langmuir.9b00154

Terzyk, A. P., Further Insights Into the Role of Carbon Surface Functionalities in the Mechanism of Phenol Adsorption, *Journal of Colloid and Interface Science*, vol. 268, no. 2, pp. 301–329, 2003. DOI: 10.1016/S0021-9797(03)00690-8

Van Erp, T. S. and Martens, J. A., A Standardization for BET Fitting of Adsorption Isotherms, *Microporous and Mesoporous Materials*, vol. 145, no. 1–3, pp. 188–193, 2011. DOI: 10.1016/J.MICROMESO.2011.05.022

Vanmoerkerke, W., Svensson, R. and Grönbeck, H., Efficient Parameterization of Adsorbate–Adsorbate Interactions on Metal Surfaces, *Surface Science*, vol. 754, p. 122678, 2025. DOI: 10.1016/J.SUSC.2024.122678

Vigdorowitsch, M., Pchelintsev, A., Tsygankova, L., and Tanygina, E., Freundlich Isotherm: An Adsorption Model Complete Framework, *Applied Sciences (Switzerland)*, vol. 11, no. 17, p. 8078, 2021.

Vijayaraghavan, K. and Yun, Y. S., Bacterial Biosorbents and Biosorption, *Biotechnology Advances*, vol. 26, no. 3, pp. 266–291, 2008. DOI: 10.1016/J.BIOTECHADV.2008.02.002

Wang, J. and Guo, X., Adsorption Isotherm Models: Classification, Physical Meaning, Application and Solving Method, *Chemosphere*, vol. 258, p. 127279, 2020. DOI: 10.1016/J.CHEMOSPHERE.2020.127279

Wang, X., Cheng, H., Ye, G., Fan, J., Yao, F., Wang, Y., Jiao, Y., Zhu, W., Huang, H. and Ye, D., Key Factors and Primary Modification Methods of Activated Carbon and Their Application in Adsorption of Carbon-Based Gases: A Review, *Chemosphere*, vol. 287, p. 131995, 2022. DOI: 10.1016/J.CHEMOSPHERE.2021.131995

Weber, W. J. and Smith, E. H., Activated Carbon Adsorption: The State of the Art, *Studies in Environmental Science*, vol. 29, no. C, pp. 455–492, 1986. DOI: 10.1016/S0166-1116(08)70958-0

Xu, W., Liu, C., Zhu, J. M., Bu, H., Tong, H., Chen, M., Tan, D., Gao, T. and Liu, Y., Adsorption of Cadmium on Clay-Organic Associations in Different PH Solutions: The Effect of Amphoteric Organic Matter, *Ecotoxicology and Environmental Safety*, vol. 236, p. 113509, 2022. DOI: 10.1016/J.ECOENV.2022.113509

Yan, H., Yang, H., Li, A. and Cheng, R., PH-Tunable Surface Charge of Chitosan/Graphene Oxide Composite Adsorbent for Efficient Removal of Multiple Pollutants From Water, *Chemical Engineering Journal*, vol. 284, pp. 1397–1405, 2016. DOI: 10.1016/J.CEJ.2015.06.030

Yang, C. Hai, Statistical Mechanical Study on the Freundlich Isotherm Equation, *Journal of Colloid and Interface Science*, vol. 208, no. 2, pp. 379–387, 1998. DOI: 10.1006/JCIS.1998.5843

Zhang, G., Yu, G., Li, D. and Wang, Y., Optimization of Inlet and Outlet Parameters of Circular Recirculating Aquaculture Tank Based on CFD-ML-MOGA, *Journal of Environmental Chemical Engineering*, p. 116648, 2025. https://linkinghub.elsevier.com/retrieve/pii/S2213343725013442. DOI: 10.1016/J.JECE.2025.116648

Zhang, X., Zhang, K., Shi, Y., Xiang, H., Yang, W. and Zhao, F., Surface Engineering of Multifunctional Nanostructured Adsorbents for Enhanced Wastewater Treatment: A Review, *Science of the Total Environment*, vol. 920, p. 170951, 2024. DOI: 10.1016/J.SCITOTENV.2024.170951

Zheng, C., Wu, Q., Hu, X., Wang, Y., Chen, Y., Zhang, S. and Zheng, H., Adsorption Behavior of Heavy Metal Ions on a Polymer-Immobilized Amphoteric Biosorbent: Surface Interaction Assessment, *Journal of Hazardous Materials*, vol. 403, p. 123801, 2021. DOI: 10.1016/J.JHAZMAT.2020.123801

6 Novel Technologies for Preparation of Adsorbent Materials

Tapas Roy, Sheetal N. Mane, Purnima Kumari, and Anupama Singh

6.1 ADSORBENT MATERIALS IN FOOD

In the food processing industry, adsorbent technology is one of the most important technologies that is used in water purification, food packaging, and storage to maintain the quality, freshness, and safety of food products (Chong et al., 2010). Adsorbent materials work by adsorbing (binding) unwanted substances such as moisture, odors, ethylene gas, and certain contaminants, thereby slowing down spoilage, preventing mold growth, and extending the shelf life of food items. Generally, types of food adsorbents are silica gel (Cheng et al., 2019), activated carbon (Mochizuki et al., 2016), bentonite clay (Alshameri et al., 2018), and zeolites (Liu et al., 2018). These materials are widely used in various food packaging solutions, such as sachets, strips, or integrated within packaging films, to protect perishable foods like fruits, vegetables, meats, and baked goods from degradation and to ensure they remain fresh for longer periods. Additionally, adsorbent materials play a vital role in the food industry, offering innovative solutions to enhance food quality, safety, and shelf life.

Adsorption is a complex surface phenomenon and a complex process that is based on the physical interactive force between two different surface polarities and molecule attraction, like the adhesion of atoms, ions, or molecules from a liquid, gas, or dissolved solid to a surface, forming a film or layer (Bruch et al., 2007; Rouquerol et al., 2013). Adsorbent material can attract and bind specific molecules or substances onto its surface through adsorption. In the adsorption process, molecules adhere to the surface of the adsorbent without penetrating. This process is crucial in various industrial, environmental, and scientific applications (Saleh et al., 2024).

For example, silica gel packets are kept in dried food products to adsorb moisture and activated carbon is used in purifying ingredients. Adsorbent materials protect and preserve the sensory qualities of food. In addition, they contribute to food safety by minimizing the growth of microorganisms and reducing the risk of contamination (Serrano-Gómez et al., 2015). Moreover, the commercial applications of adsorbent technologies in the food industry are diverse and integral to

DOI: 10.1201/9781003767251-6

preserving food quality and safety across the supply chain. New advancements continue to enhance the effectiveness and sustainability of these technologies, making them increasingly valuable in modern food packaging and storage solutions. As per the demand for maximum retention of nutritional properties and increased shelf-life duration, the application of adsorbent materials continues to expand, proving to be an essential component in the food industry to meet consumer expectations (Wu et al., 2022).

Factors affecting the effectiveness of adsorbent material:

- **Surface Area and Porosity**
  - **High Surface Area**: Adsorbent materials usually have a large surface area to interact with the target substance. The greater surface area provides more space for molecules to adhere.
  - **Porous Structure**: Many adsorbents, such as activated carbon and zeolites, have tiny pores that increase their surface area and make them highly effective at capturing molecules. These pores can vary in size, allowing adsorbents to selectively attract molecules based on size or shape.
- **Physical Adsorption (Physisorption)**
  - **van der Waals Forces**: This type of adsorption occurs due to weak intermolecular forces, known as van der Waals forces, where molecules from a gas or liquid phase adhere to the surface of the adsorbent. This process is generally reversible, meaning the adsorbed substances can be desorbed or released by changing conditions like temperature or pressure.
  - **Non-Specific Binding**: Physisorption does not involve a chemical reaction and can attract a wide variety of molecules, making it versatile for different applications such as moisture control or odor removal.
- **Chemical Adsorption (Chemisorption)**
  - **Chemical Bonds**: In chemisorption, molecules form stronger chemical bonds with the adsorbent surface. This process is usually selective and involves a chemical reaction, such as oxidation or reduction.
  - **Specificity**: Because chemisorption involves specific chemical interactions, it is more targeted than physisorption and is used for applications requiring the selective removal of certain compounds, such as purifying gases or removing contaminants.

6.2 TYPES OF ADSORBENT MATERIALS AND ITS PURPOSE

6.2.1 Silica Gel

Silica gel is a widely used adsorbent in the food industry, primarily known for its moisture-absorbing properties. It is a porous, granular material derived from silicon dioxide (SiO_2), and it comes in a variety of forms, including beads, packets, and sachets (Figures 6.1 and 6.2). Silica gel has a high surface area with a network of tiny pores, which allows it to adsorb large amounts of moisture relative to its weight. It attracts and holds water molecules from the surrounding environment through adsorption, which helps control humidity levels. It is generally

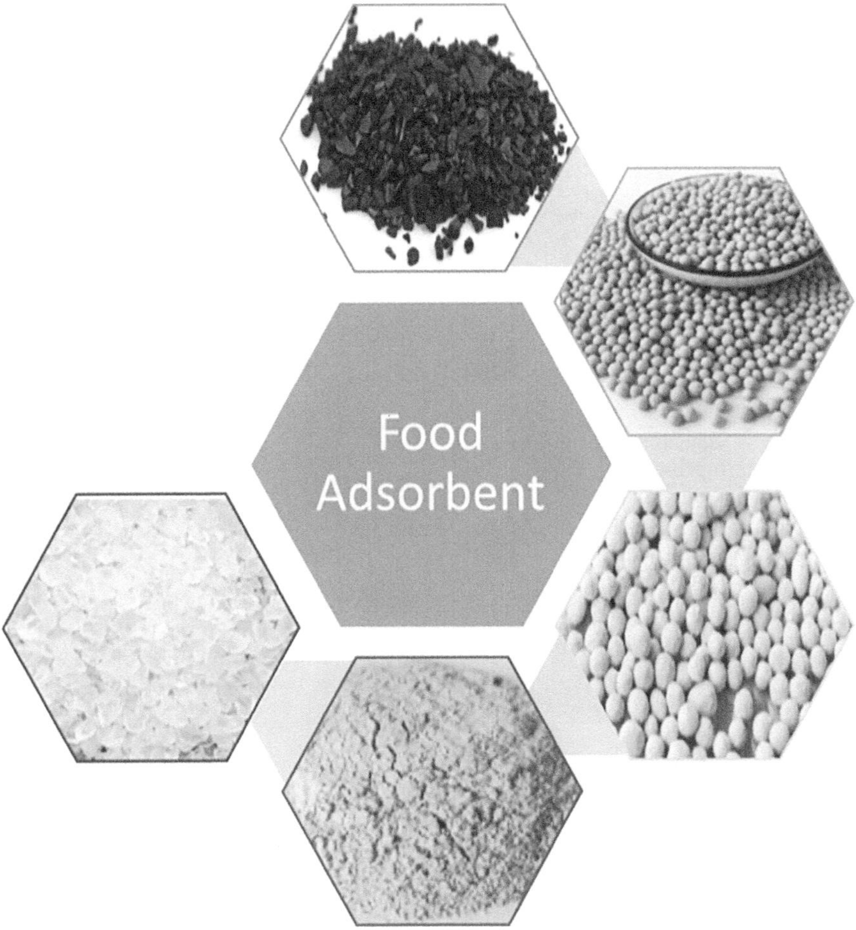

FIGURE 6.1 Different types of food adsorbent materials.

considered safe for use in food packaging, provided it is food grade and free of contaminants.

6.2.2 Activated Carbon

Activated carbon, also known as activated charcoal, is a highly porous material derived from carbonaceous substances like coal, wood, or coconut shells. It is treated to develop a large surface area with extensive porosity, making it an effective adsorbent for a wide range of substances. Activated carbon has a large surface area with numerous tiny pores, which enhances its adsorption capacity. The porous structure allows activated carbon to trap molecules of gases, liquids, and contaminants. Activated carbon works through physical adsorption, where molecules adhere to the surface of the material due to van der Waals forces.

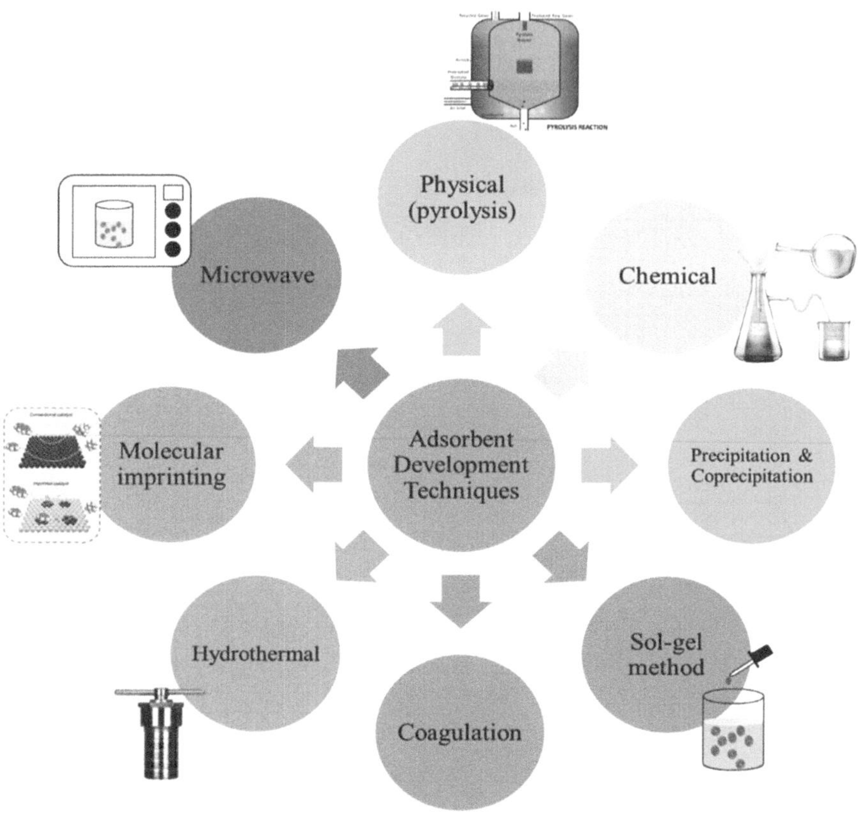

FIGURE 6.2 Different techniques for the development of adsorbent material.

6.2.3 Zeolites

Zeolites are a group of crystalline aluminosilicates with a porous structure. Their unique properties make them highly effective adsorbents for a variety of applications. The framework of a zeolite is composed of interconnected tetrahedral units of silicon and aluminum atoms, each surrounded by four oxygen atoms. The arrangement of these tetrahedral units creates a three-dimensional network of pores, with a specific size and shape. Zeolites can exchange their cations (e.g., sodium, potassium) with other ions present in the surrounding environment.

6.2.4 Activated Alumina

Activated alumina is a highly porous form of aluminum oxide (Al_2O_3). Its structure consists of a network of interconnected pores, which provide a large surface area for adsorption. The extensive pore network in activated alumina provides a vast surface area, allowing it to adsorb a wide range of substances. Activated alumina is chemically stable and resistant to most acids and bases, making it suitable for various applications. It can withstand high temperatures without losing its adsorptive properties. Activated alumina can selectively adsorb certain substances based on their molecular size and polarity.

6.2.5 Bentonite Clay

Bentonite clay is a hydrated aluminum silicate (Al_2SiO_5) mineral with a high swelling capacity. It is primarily composed of montmorillonite, a clay mineral known for its ability to absorb water and expand significantly. It can absorb large amounts of water, causing it to swell and increase in volume. This property is due to the presence of interlayer cations, such as sodium or calcium, which are attracted to water molecules. Bentonite clay has colloidal properties, showing it forms a stable suspension in water. It exhibits thixotropic behavior, resulting in it becoming less viscous when agitated. Bentonite clay can adsorb various substances, including heavy metals, organic compounds, and dyes.

6.3 TECHNIQUES FOR THE DEVELOPMENT OF ADSORBENT MATERIALS

The development of adsorbent materials is crucial for various applications, such as water purification, air filtration, and industrial waste treatment. The techniques for developing these materials can be broadly categorized into conventional, green, and advanced techniques.

6.3.1 Conventional Techniques

Conventional methods have been used for decades to develop adsorbent materials. When selecting an activation method for adsorbent material, several factors need to be considered, including the time required for material synthesis, the type and toxicity of the reagent used, the properties of the produced material, the cost of activation, and, most importantly, the complexity of the process and the equipment required (Ahmed et al. 2019). These methods have been discussed in detail as follows.

6.3.1.1 Physical Activation

The physical activation process, also referred to as gas activation, involves pyrolysis, which primarily takes place at elevated temperatures. This process generally consists of two main steps: first, carbonizing the carbon substances in a protective environment, and then activating the resulting substance using gasifying agents, namely air, steam, CO_2, or a mixture of these. Generally, the activation temperatures vary between 700°C and 1100°C. The following equations are an illustration of the gas–water reaction:

$$C + H_2O = CO + H_2 \tag{6.1}$$

$$C + 2H_2O = CO_2 + 2H_2 \tag{6.2}$$

The carbon is gasified during this process, forming porous, highly activated carbon. Physical activation is noted for its minimal environmental impact. Specifically, when using CO_2 and steam as activation agents, there is a significant rise in the specific surface area: activation with CO_2 enhances the surface area by about 89–653 m^2/g. In contrast, activation with steam increases it from 89 to 1015 m^2/g (Mai et al. 2021). Rashed et al. (2018) experimented with activated sewage sludge heated to 800°C to effectively remove Cd, Cu, and Pb from effluent. The optimum removal parameters were pH 6, room temperature, and a 60-minute duration. According to the findings, the removal efficiencies were 99.5% (Cd) and 97.7% (Pb), following the Freundlich model, and 93.7% for Cu, following the Langmuir model. Similarly, Ibrahim et al. (2016) conducted an experiment on the elimination of toxic compounds (Cr, Cu, Pb, and Cd) of sewage utilizing the macroalgae *Ulva lactuca*. The algae were activated at 800°C for 3 hours and then used in batch adsorption tests for Cr, Cu, Pb, and Cd. The adsorption limits achieved were 82 mg/g (Cr), 84.7 mg/g (Cu), 83.3 mg/g (Pb), and 84.6 mg/g (Cd) at a pH of 5, with an equilibrium time of 1 hour, as modeled by the Langmuir isotherm. Tsai and Jiang (2018) investigated activated carbons made from dried shells of coconut by applying a one-step physical activation approach. This process involved carbonization at 500°C under nitrogen, followed by activation with CO_2 at temperatures between 700°C and 900°C. The study found that increasing the activation temperature and retention time enhanced pore formation because of the intense reaction between CO_2 and lignocellulose-based carbon. The resulting ACs achieved a surface area of 1100 m^2/g and a mesopore rate exceeding 40%. The benefits of this approach include eliminating contaminants typically introduced by activating agents, eliminating the necessity of post-synthesis purification, being cost-effective, and employing a non-corrosive process. However, potential drawbacks include a prolonged synthesis period, elevated pressure and temperature use, and minimal ability to regulate permeability (Hernández-Montoya et al., 2012).

6.3.1.2 Chemical Activation

The chemical activation approach involves impregnating the raw source of carbon with appropriate chemicals in a single-step procedure. Common activation agents include salts based on alkali metals and other acids or bases (e.g., HCl, HNO_3, H_3PO_4, $ZnCl_2$, H_2SO_4, NaOH, and KOH), and subsequently carbonization at reduced temperatures. This approach improves the pore volume and surface area. Additionally, this method offers benefits like scalability, reduced energy and consumption, and effective pore development. However, the purification process can be challenging due to the need to remove residual impurities, and using corrosive and hazardous chemicals poses potential waste disposal risks. Activated carbon developed via chemical activation using the golden shower tree blooms showed a significant surface area of 1413 m^2/g (Tran et al., 2018). In contrast, the chemical activation of *Eichhornia* (water hyacinth) using KOH resulted in a surface area of 900 m^2/g and an average pore diameter of 21.6 Å (Salman et al., 2021).

In one more research study by Gebretsadik et al. (2020), the detoxification of heavy metals (Pb, Cd, and Cr) through aqueous solutions utilizing Eucalyptus camaldulensis was investigated. Adsorption data for Cd and Pb followed the Freundlich and pseudo-second-order models, with adsorption capacities of 49.5 mg/g and 102 mg/g, respectively. Optimal adsorption occurred at pH 7, with an adsorbent dosage of 50 mg at 30°C. Additionally, $ZnCl_2$ was utilized to create activated carbon from teak sawdust by hydrothermal/chemical activation, achieving adsorption capacities of 208 mg/g for Cd and 182 mg/g for Cu, according to the Langmuir model. It was found that oxygen functional groups significantly contributed to the adsorption capacity, with complexation mechanisms proving more effective than electrostatic interactions. Furthermore, Eucalyptus camaldulensis was chemically activated using H_2SO_4 to remove Pb, Cd, and Cr from sewage in batch adsorption trials. The data aligned with the Langmuir model, showing adsorption capacities of 47.62 mg/L (Cr), 6.17 mg/L (Cd), and 9.26 mg/L (Pb) in ordinary solutions (pH 10–12), indicating a single-layer coverage adsorption mechanism. The carbon material derived from left-over banana peel using 0.5 M H_2SO_4, by chemical activation, proved effective in removing zinc, lead, and chromium ions. The removal efficiency was influenced by factors, namely, adsorbent dosage, particle size, contact time, and pH. It was also noted that the removal efficiency of lead (Pb) and chromium (Cr) ions was reduced with increasing particle size and contact period (Abdulfatai et al. 2013). Castro et al. (2022) synthesized an adsorbent utilizing coffee husk (a byproduct of the coffee industry) with the help of zinc chloride and phosphoric acid. The adsorbent, activated with phosphoric acid, exhibited a porous structure with a high surface area of 429.7 m^2/g. Its effectiveness was assessed for treating effluents comprising synthetic Ponceau 4R dye, and the process was found to be thermodynamically favorable, as indicated by the negative Gibbs free energy and a positive entropy value of +0.022 kJ mol^{-1} K^{-1}.

6.3.1.3 Precipitation and Co-Precipitation

In the precipitation process, metal salts are dissolved in a solution, followed by adding a precipitating agent, forming solid particles. These particles are filtered, washed, and calcined to obtain the adsorbent material. Precipitation can be done employing lime, soda ash, and sodium sulfide. Chemical precipitation offers various advantages, including an easy process that can be done at ambient conditions, in various pH levels, and is unaffected by low heavy metal ion concentrations. Nevertheless, substantial chemical doses are required, and treating sludge induced by chemical precipitation raises the expense (Wang et al. 2023). Precipitation involves the formation of insoluble metal as a sediment, and it is among the processes of metabolism-dependent biosorption (Table 6.1). However, metabolism-independent biosorption can also arise through precipitation. According to Perpetuo et al. (2011), precipitation happens in metabolism-dependent biosorption as a result of microorganisms activating a defense mechanism in response to exposure to hazardous metal ions. Whereas, in metabolism-independent biosorption, chemical interactions between the metal ions and the functional groups on the biosorbent's cell wall potentially involve redox reactions to cause precipitation. The primary mechanisms for the reduction of Fe^{3+} ions in an aqueous solution using green tomato husk were precipitation and ion exchange. At pH 6.0, Fe^{3+} ions were precipitated as iron hydroxide (FeOOH) (García-Mendieta et al., 2012)

TABLE 6.1
Overview of Innovative Methods for Adsorbent Material Synthesis

Technique	Description	Advantages	Disadvantages	Reference
Physical activation	Involves carbonization followed by activation employing gases like CO_2 or steam	High surface area and porosity	Energy-intensive and requires high temperatures	Bachrun et al. (2016)
Chemical activation	Carbon material treated with chemical agents like KOH or $ZnCl_2$ at lower temperatures to increase porosity	High yield and more controlled pore structure	Chemical handling can be hazardous, and washing to remove chemicals is required	Gong (2013)
Sol–Gel method	A wet-chemical process that involves the conversion of a liquid ("sol") (a colloidal dispersion of particles) to a solid "gel" state. It is widely used for creating porous adsorbent materials	- High control over porosity and surface area - Can synthesize diverse materials	- Requires a long processing time - Expensive precursors	Almeida and Gonçalves (2021)
Precipitation and co-precipitation	Formation of solid adsorbents from a solution by adding precipitating agents, often used for metal oxides	Simple, cost-effective, and widely applicable	Limited control over particle size and structure	Pu et al. (2018)
Bioleaching	Utilizes microorganisms to leach metals from ores, producing biobased adsorbents	Eco-friendly and low energy requirement	Slow process and limited applicability to certain materials	Camargo et al. (2018)
Coagulation/ flocculation	Involves adding coagulants to wastewater, forming aggregates that can be separated easily	Effective for water treatment and low cost	Generates sludge that requires disposal	Iwuozor (2019)
Green synthesis	Uses biological materials (e.g., plant extracts) to synthesize adsorbents, offering an eco-friendly and sustainable approach	- Eco-friendly and sustainable - Reduces use of toxic chemicals	- Lower control over material properties - Limited reproducibility	Sharma et al. (2022)
Activated carbon		High Efficiency, High adsorption capacity, versatile application, effective against various contaminants, and readily available commercially	Performance relies on the adsorbent type, expensive, no regeneration	Li et al. (2020)

Biochar	Charcoal produced from biomass in the absence of oxygen, used as a sustainable adsorbent	Renewable, cost-effective, and easily scalable	Limited adsorption capacity compared to advanced materials	Singh et al. (2024)
Nanomaterials (carbon nanotubes)	Utilizing nanoscale materials for the remediation of polluted water	Highly selective and adsorption capacity	Expensive, no regeneration capacity	Wei et al. (2019)
Plant-based adsorbent (agricultural waste)	Uses plant biomass (e.g., leaves, seeds) as natural adsorbents for pollutants	Economical, relatively higher metal removal capacities, and the most significant potential for metal removal, recyclable	More solid waste, limited research, and application	Al-Qodah and Al-Shannag (2017), Al-Qodah et al. (2017)
Algae-based adsorbent	Uses algae biomass to develop natural adsorbents, particularly effective for heavy metal removal	Abundant, renewable, and biodegradable	Variable adsorption efficiency depending on algae species	Upadhyay et al. (2024)
Photocatalysis		Concurrent removal of metals and organic pollutants, less hazardous bioproduct	Needs extended time for separation, limited applications	Al-Qodah and Al-Shannag (2017), Al-Qodah et al. (2017)
Hydrothermal carbonization	Uses high temperature and pressure to grow crystals from aqueous solutions. Suitable for producing microporous adsorbents	- Produces high-quality crystals with controlled size and morphology - Suitable for microporous adsorbents	- Requires high temperature and pressure - Expensive to set up and energy-intensive	Alatalo and Sillanpää (2020)
Metal–organic framework (MOF)	Porous crystalline materials formed by metal ions and organic linkers, used for gas adsorption and separation	High surface area, tunable porosity, and selective adsorption	Expensive synthesis and instability in humid environments	Lin et al. (2020)
Molecularly imprinted polymers (MIPs)	Polymers with specific recognition sites for target molecules, used for selective adsorption	High selectivity and reusability	Complex synthesis and high cost of monomers	Cantarella et al. (2019)
3D printing	Additive manufacturing technique used to fabricate adsorbent materials with controlled geometries and porosity	Customizable, scalable, and efficient material use	High equipment cost and limited material options for adsorption	Jin et al. (2022)

6.3.1.4 Sol–Gel Method

This method entails the conversion of a liquid ("sol") (a colloidal dispersion of particles) to a solid "gel" state. The resulting material is then dried and heat-treated to form the adsorbent. The common materials used are silica gels (SiO_4) and alumina (AlO_4). This method is frequently used to produce ceramic and glassy materials by starting with a "sol" solution, which undergoes gelation and subsequent ejection of the liquid phase. The resulting materials can be called sonogel, alcohologel, aerogel, cryogel, or xerogel. The sol-gel method has been invented to surmount the issues related to powdered activated carbon (PAC). This method is highly stable and non-leaching. According to Sanchez (2014), the sol–gel method aids the production of particulate materials with a large surface area and excellent mechanical characteristics, which are desirable traits for an adsorbent. The synthesized adsorbent composites using the silica precursor tetraethyl orthosilicate (TEOS) and chitosan. These composites exhibited improved chemical and physical properties, particularly with large surface areas ranging from 374.94 m^2/g to 886.31 m^2/g. Similarly, Chintokoma (2024) synthesized a composite adsorbent coating from Prosopis juliflora-activated carbon using the sol–gel method to adsorb Cd^{2+} and $Cr_2O_7^{2-}$ from wastewater. Under optimized parameters (pH 8.5, 5 mg concentration, 0.25 g dosage, 105 minutes contact time, and 23.73°C temperature), the removal efficiencies of the metal ions were found to be 83.98% for Cd^{2+} and 58.08% for $Cr_2O_7^{2-}$. Furthermore, composite adsorbent coatings were proven effective substitutes for PAC, providing considerable benefits in metal ions elimination from wastewater.

6.3.1.5 Bioleaching

Li and Xiao (2019) produced an activated carbon of rice husk char using K_2CO_3 leaching after CO_2 activation. It featured a dual pore structure with a specific surface area of 1097 m^2/g, micropore volume of 0.34 cm^3/g, and mesopore volume of 0.49 cm^3/g. The activated carbon showed an excellent CO_2 adsorption capacity (3.1 mmol/g), selectivity (7.6 at 25°C), fast adsorption kinetics, and good recyclability, comparable to KOH-activated carbons, indicating a potentially suitable sorbent for CO_2 capture.

6.3.1.6 Coagulation/Flocculation

Coagulation/Flocculation is a physicochemical method used to decrease suspended solids and colloidal turbidity (Gautam and Saini, 2020). Coagulation is a process in which pollutants and suspended particles collide with oppositely charged particles, causing them to aggregate and form an insoluble complex that eventually settles as sediment (Syam Babu et al., 2020). The key factors influencing contaminant removal are the settling time, coagulant dosage, and pH.

6.3.1.6.1 Traditional Adsorbent Synthesis: Challenges and Limitations

Traditional methods for adsorbent synthesis typically involve chemical activation processes that use acids, bases, or salts to improve surface area and pore formation in the materials. These methods, while effective, often require harsh chemicals and high temperatures, resulting in environmental pollution and high production costs.

Moreover, conventional adsorbents may lack specificity for emerging contaminants such as pharmaceuticals, dyes, and microplastics, necessitating the development of novel adsorbent materials with improved selectivity and performance.

6.3.2 Green Techniques for Adsorbent Synthesis

Green synthesis approaches prioritize sustainability by using renewable resources, non-toxic reagents, and energy-efficient processes. These techniques are becoming more popular as they align with green chemistry principles, which aim to minimize the environmental impact of chemical processes.

6.3.2.1 Biomass-Derived Adsorbents

Biomass-derived adsorbents have gained considerable interest because they are renewable, cost-effective, and capable of utilizing waste materials from agriculture and industry. Biomass consists of various materials, including fruit peels, algae, wood, and agricultural residues, which are abundant in carbon and contain functional groups like hydroxyl, carboxyl, and amine groups (Egbuchunam et al., 2018). These materials can be processed into biochar or activated carbon using green synthesis methods.

6.3.2.1.1 Biochar

Biochar is produced through the pyrolysis of biomass in an oxygen-limited environment. Unlike traditional activated carbon production, which often involves the use of strong acids or bases, biochar production can be achieved through environmentally friendly methods, such as slow pyrolysis at low temperatures (300°C–600°C) without harmful chemicals. Biochar has a porous structure, a vast surface area, and various functional groups that enable it to adsorb heavy metals and organic pollutants effectively (Akhtar et al., 2024).

Biochar derived from various agricultural byproducts, including corn cobs, rice husks, and sawdust, exhibits high adsorption capacities for heavy metals such as copper (Cu), lead (Pb), and cadmium (Cd). These materials offer a sustainable solution to both waste management and environmental remediation. For example, Rice husk is a widely available agricultural waste material that has been converted into biochar through pyrolysis under low temperatures. The resulting biochar exhibits high porosity and surface area, rendering it a potent adsorbent for heavy metals like mercury and lead (Shen et al., 2021). The green synthesis of rice husk biochar minimizes energy consumption and reduces greenhouse gas emissions compared to traditional methods.

6.3.2.1.2 Algae-Based Adsorbents

Algae, particularly macroalgae and microalgae, offer a promising renewable resource for adsorbent synthesis due to their fast growth rates and high biomass yields. Algal cell walls consist of polysaccharides, proteins, and lipids, which enhance the ability to adsorb heavy metals and dyes. Green methods for algae-based adsorbents typically involve harvesting, drying, and minimal chemical activation of algae. One such method is the pyrolysis of algae into algal biochar under mild conditions, which can be used for water treatment.

For example: *Spirulina-Based Adsorbents*: Spirulina, a blue-green microalga, has been utilized to absorb heavy metals like copper, cadmium, and zinc. The adsorption process occurs through metal ion exchange with functional groups present on the algal cell membrane. Spirulina-based bioadsorbents have shown significant promise for the removal of heavy metals like arsenic and lead from polluted water sources. Furthermore, the green synthesis of algae-based bioadsorbents does not generate secondary pollutants, making it a clean and renewable solution for water remediation (Antonopoulou et al., 2024).

6.3.2.1.2.1 Seaweed-Derived Adsorbents Macroalgae, such as seaweed, have been converted into biochar and activated carbon for water treatment applications. The alginate in seaweed acts as a natural adsorbent for cationic pollutants, offering an environmentally friendly alternative to synthetic adsorbents (Devrajani et al., 2024).

6.3.2.2 Plant-Based Green Synthesis

Plant materials, including leaves, fruit peels, and seeds, provide a sustainable and eco-friendly source for adsorbent synthesis. These materials contain natural polymers, including cellulose, hemicellulose, lignin, and pectin, which can be altered via physical or chemical activation to improve their adsorption capabilities. Plant-based materials are abundant, renewable, and biodegradable, which makes them attractive for green synthesis approaches (Wang et al., 2013).

6.3.2.2.1 Orange Peel Adsorbent

Citrus fruit peels, especially orange peels, are rich in pectin, which contains hydroxyl and carboxyl groups that can interact with pollutants. Through green synthesis processes such as low-temperature pyrolysis and steam activation, orange peels can be transformed into bioadsorbents for removing metal ions, dyes, and organic pollutants from wastewater. Researchers have found that orange peel adsorbents are effective in removing cadmium and chromium from wastewater, with adsorption capacities of up to 95% for some metals. The process requires minimal chemical inputs, making it an eco-friendly alternative to conventional adsorbent production (Akinhanmi et al., 2020).

6.3.2.2.2 Moringa Oleifera Seeds

Moringa oleifera seeds are another plant-based material with significant adsorptive potential. The seed extract contains proteins and peptides that act as natural coagulants and flocculants for water purification. Green synthesis methods involve the direct use of Moringa seed powder or its protein extract without chemical modification, which effectively removes turbidity and contaminant metals from water (Vieira et al., 2010).

6.3.2.3 Hydrothermal Carbonization (HTC)

HTC) is an environmentally friendly technique employed to transform biomass into adsorbent material at relatively low temperatures (180°C–250°C) and under subcritical water conditions (Owsianiak et al., 2016). Unlike traditional pyrolysis, HTC occurs in a water medium, eliminating the need for harsh chemicals and reducing energy consumption. HTC can produce carbon-rich adsorbents with high porosity and oxygen-containing functional groups, making them suitable for the elimination of both organic and inorganic contaminants from water.

A significant advantage of HTC is its ability to process wet biomass without predrying, which makes it especially appropriate for materials such as algae, food scraps, and sewage sludge. HTC has been used to synthesize adsorbents from agricultural waste, such as corncob hydrochar, which effectively removes dyes and heavy metals from wastewater.

6.3.2.4 Green Synthesis of Nanomaterials

Nanotechnology has introduced novel methods for adsorbent synthesis, particularly through the green synthesis of nanoparticles. These methods utilize plant extracts, algae, or microorganisms as reducing and stabilizing agents, offering a sustainable alternative to traditional chemical routes that involve toxic solvents or high energy requirements (Aboagye et al., 2024; Prathna, 2019).

6.3.2.4.1 Green Synthesis of Metal Nanoparticles

Metal nanoparticles like gold, silver, and iron oxide nanoparticles (IONPs) are increasingly being used in adsorption applications because of their large surface area and active sites. Green synthesis of metal nanoparticles using plant extracts has been widely researched. For instance, *Azadirachta indica* (neem) leaves have been used to synthesize silver nanoparticles with high efficiency, which are capable of binding to heavy metals and organic dyes.

The plant extract acts as a bioreductant, converting metal ions into nanoparticles without the need for hazardous chemicals. This eco-friendly process reduces energy consumption and generates fewer byproducts, making it a promising approach for sustainable adsorbent synthesis.

6.3.2.4.2 Iron Oxide Nanoparticles From Green Methods

IONPs synthesized via green routes have been shown to exhibit superior adsorption capacity for contaminant metals and anions, including arsenic and fluoride, from water. Researchers have used plant extracts from tea leaves and eucalyptus to synthesize IONPs, resulting in nanoparticles with high surface area and reactive sites that can remove pollutants through ion exchange and complexation.

6.3.3 Advanced/Novel Techniques in Adsorbent Synthesis

Besides green methods, a range of innovative techniques has been introduced to produce advanced adsorbent materials that offer improved performance and selectivity.

6.3.3.1 Metal–Organic Frameworks (MOFs)

MOFs are crystalline materials with porous structures, formed by the coordination of metal ions with organic ligands. They are distinguished by their large surface areas, adjustable pore configurations, and diverse chemical properties, which enhance their effectiveness as adsorbents for applications in gas storage, separation, and environmental remediation (Li & Yang, 2015).

6.3.3.1.1 Synthesis of MOFs

MOFs are produced through solvothermal or hydrothermal techniques. Solvothermal synthesis involves heating a mixture of metal ions and organic linkers in a solvent at elevated temperatures, leading to the formation of a highly porous network (Avci & Erucar, 2018). However, recent innovations in green synthesis have focused on reducing the use of organic solvents, replacing them with water or bioderived solvents, and utilizing lower temperatures and pressures to reduce energy consumption.

6.3.3.1.2 Applications of MOFs

MOFs are especially efficient at adsorbing gases such as carbon dioxide (CO_2), methane (CH_4), and volatile organic compounds (VOCs). For instance, MOFs such as HKUST1 and MIL101 have shown high CO_2 adsorption capacities and selectivity, making them ideal for carbon capture applications. Additionally, MOFs can be functionalized with specific ligands to enhance their affinity to heavy metals and organic contaminants in wastewater. MOF5, a well-known MOF, has been thoroughly researched for its capacity to adsorb gases like carbon dioxide and methane. Its large surface area and pore volume render MOF5 particularly suitable for applications in gas storage and separation. Novel green synthesis methods for MOFs, such as mechanochemical and microwave-assisted techniques, reduce the energy and solvent requirements typically associated with MOF production. Zeolitic imidazolate frameworks (ZIFs), a subclass of MOFs, have shown great potential for adsorbing organic pollutants from water. ZIF8, in particular, has been used for removing dyes and pesticides due to its hydrophobic pore environment. Recent studies have focused on green synthesis routes for ZIF8 using aqueous solutions and mild reaction conditions.

6.3.3.2 Covalent Organic Frameworks (COFs)

COFs represent another category of highly porous materials synthesized from organic molecules linked by covalent bonds. COFs exhibit crystalline structures, high surface areas, and tunable pore sizes, which allow for selective adsorption of specific contaminants. Compared to MOFs, COFs are composed entirely of lightweight elements (C, H, N, O, and B), offering advantages in terms of stability and low density.

6.3.3.2.1 Synthesis of COFs

COFs are typically synthesized via a reversible condensation reaction between organic monomers. One of the challenges in COF synthesis is ensuring crystallinity while maintaining high porosity. Recent advancements in mechanochemical synthesis, which involves grinding reactants together without solvents, have enabled the production of COFs with minimal environmental impact.

6.3.3.2.2 Applications of COFs

COFs are particularly promising for the removal of organic pollutants, such as dyes and pharmaceuticals, from water. For example, COF-based adsorbents have been designed with functionalized pores that selectively capture antibiotic molecules from wastewater. Additionally, COFs have shown potential in gas adsorption applications, particularly for hydrogen storage and carbon capture. COFs have been developed for carbon capture applications due to their high surface areas and tunable pore sizes. By incorporating amine groups into the framework, COFs can selectively adsorb CO_2 from gas mixtures, offering a potential solution for mitigating greenhouse gas emissions. The use of green chemistry principles in COF synthesis, such as solvent-free reactions, has improved the sustainability of these materials.

6.3.3.3 Molecularly Imprinted Polymers (MIPs)

MIPs are synthetically created polymers designed with specific recognition sites tailored for target molecules. These polymers are produced by polymerizing functional monomers around a template molecule, which is the target pollutant. After the template is extracted, the polymer matrix holds imprinted cavities that match the template in size, shape, and function, enabling selective adsorption. (Ballesteros-Gómez et al., 2009).

6.3.3.3.1 Synthesis of MIPs

MIPs are typically synthesized using free radical polymerization, with crosslinking agents used to stabilize the polymer matrix. Green synthesis methods for MIPs have focused on reducing the use of toxic solvents during polymerization and incorporating biobased monomers and crosslinkers.

6.3.3.3.2 Applications of MIPs

MIPs have proven effective for the targeted extraction of pharmaceuticals, pesticides, and heavy metals from wastewater. For instance, MIPs designed to target antibiotics such as tetracycline have demonstrated high selectivity and adsorption capacities in water treatment applications.

6.3.3.4 3D Printing and Additive Manufacturing of Adsorbents

Additive manufacturing (3D printing) is an emerging technology that offers the potential to design and fabricate adsorbents with precisely controlled pore structures, shapes, and functionalities. This technique allows researchers to tailor the physical and chemical characteristics of adsorbents for targeted applications, including water treatment, air purification, and gas separation.

6.3.3.4.1 Materials for 3D-Printed Adsorbents

3D-printed adsorbents can be created using a range of materials, such as polymers, ceramics, and composites. In particular, polymer-based adsorbents embedded with nanoparticles (such as graphene or metal oxides) offer enhanced adsorption performance for removing organic and inorganic pollutants. Additive manufacturing allows for the design of hierarchical pore structures that improve mass transfer and adsorption efficiency.

6.3.3.4.2 Applications of 3D-Printed Adsorbents

3D-printed adsorbents have been applied in various environmental and industrial sectors, including water treatment, air purification, and catalysis. The ability to customize pore size and shape enables the creation of adsorbents that can target specific pollutants with high efficiency. For example, 3D-printed MOF structures have been designed for CO_2 capture in gas separation processes.

6.3.3.5 Hierarchically Structured Porous Materials

Hierarchically structured adsorbents feature pores of different sizes (micro-, meso-, and macropores), allowing for enhanced mass transfer and improved adsorption performance. These materials can be synthesized using templating methods, where a sacrificial template is used to create the desired pore structure.

6.3.3.5.1 Example: Mesoporous Silica for Heavy Metal Removal

Mesoporous silica materials, such as SBA15 and MCM41, are widely used in adsorption applications because of their well-ordered porous structures and large surface areas. By functionalizing the surface of mesoporous silica with amine groups, researchers have created adsorbents that selectively capture heavy metals like mercury and arsenic from water. Green synthesis methods for mesoporous silica involve using biobased templates, such as plant extracts, to reduce the environmental impact of the production process.

6.3.3.6 Nanomaterials for Adsorption

Nanomaterials, including metal oxide nanoparticles, graphene, and carbon nanotubes (CNTs), have garnered significant attention for their exceptional adsorption capabilities due to their high surface area to volume ratios and unique electronic properties.

6.3.3.6.1 Examples: Graphene Oxide for Dye Removal

GO, a form of graphene, has been extensively researched for its capacity to adsorb contaminant metals and dyes from water. The functional groups containing oxygen on GO create active sites that facilitate the binding of pollutants. Green synthesis methods for GO involve the use of plant extracts or biopolymers as reducing agents, eliminating the need for hazardous chemicals (Zhao & Guo, 2012).

6.3.3.6.2 Iron Oxide Nanoparticles for Arsenic Removal

IONPs are highly effective for adsorbing arsenic due to their strong affinity for arsenate ions. Novel synthesis techniques, such as green synthesis using plant extracts, have produced IONPs with enhanced adsorption capacities and minimal environmental impact.

6.4 APPLICATIONS OF ADSORBENT MATERIALS IN THE FOOD INDUSTRY

6.4.1 Activated Carbon

Activated carbon is utilized in various sectors of the food industry due to its ability to adsorb a wide range of substances, including unwanted odors, colors, and contaminants. It is used in clarifying liquids by removing unwanted colors and flavors, resulting in improvement in the overall appearance and taste of food products. In the production of beverages like vodka, whiskey, or wine, activated carbon is used to remove colorants and impurities that may affect the flavor and clarity. It is also used in filtration systems during the processing of juices, syrups, and other liquid foods. In addition, activated carbon is used to eliminate unwanted odors from food products and their packaging. Activated carbon can be incorporated into packaging materials or used in sachets inside packages to maintain a neutral smell. Activated carbon filters are commonly found in water treatment plants and in-house filtration systems to ensure high-quality water for processing foods and beverages. It is used to remove impurities and contaminants from edible oils, enhancing their quality and safety. In the refining of cooking oils, activated carbon can adsorb free fatty acids, colorants, and other unwanted substances, resulting in a cleaner and more stable product.to control moisture and gas levels, extending the shelf life of food products. To extend the shelf life of perishable items by controlling moisture and gas levels, activated carbon can be integrated into packaging materials and can also be incorporated into modified atmosphere packaging (MAP) to absorb excess gases and maintain optimal conditions for perishable items such as meat, cheese, and baked goods. These adsorbents are used for air purification purposes, namely to control odors and pollutants in food processing and storage facilities. In large-scale food processing plants, activated carbon is employed as an air filtration system to remove VOCs and other contaminants from the outgoing air, ensuring a clean environment for food production.

Activated carbon can adsorb solvents and chemicals used in food processing, allowing for their recovery and reuse. In the production of certain food additives and flavorings, activated carbon can capture and recycle solvents, reducing waste and improving process efficiency. Activated carbon is sometimes used in health foods and dietary supplements for its purported detoxifying properties. Activated carbon tablets or capsules are marketed as dietary supplements for digestive health, although scientific evidence supporting these claims is limited.

6.4.2 Zeolites

Zeolites are naturally occurring or synthetic crystalline aluminosilicates with a porous structure that can adsorb a variety of molecules. Their unique properties make them useful in several applications within the food industry.

Zeolites are used to control moisture levels in food packaging and storage. Zeolite packets are used in packaging for dried foods, grains, and powdered products to prevent moisture buildup, which can lead to spoilage, mold growth, and product degradation. Zeolites adsorb gases like ethylene, which accelerates ripening, and odors from food products. It is used in the packaging of fruits and vegetables to absorb ethylene gas and extend shelf life. It helps to maintain the freshness and quality of perishable items during transportation and storage. Zeolites are used to control odors in food storage and processing plants. Zeolites are used to soften water by exchanging calcium and magnesium ions with sodium ions. It is used in water treatment systems to purify water for food processing and production, and also prevents scale formation in equipment to improve the quality of water for various food applications.

Zeolites are used in food processing to enhance the quality of food products. It can adsorb and remove various contaminants from food and beverage processing. Moreover, it can be used to inhibit mold growth by controlling moisture levels and used in packaging for products sensitive to mold, such as bread and other baked goods. Zeolites are employed as catalysts in various chemical reactions to enhance reaction efficiency for the production of certain food additives and flavorings. These adsorbents are also used as additives in animal feed to improve feed efficiency and animal health.

6.4.3 Bentonite Clay

This type of absorbent clay, formed from volcanic ash, is used in the food industry for its unique properties, such as adsorption, decolorization, and clarification. Bentonite clay is used to clarify liquids by removing suspended particles and impurities. In the production of wines and juices, bentonite clay helps to remove proteins, yeast, and other particulates that can cause cloudiness and improve the clarity and stability of beverages, resulting in a more visually appealing and stable product. It can absorb undesirable colors and odors from food products. Also used in the processing of edible oils to remove colorants and unwanted odors, enhancing the purity and quality of the oil.

Bentonite clay can adsorb toxins and harmful substances in food processing. Used to remove mycotoxins and other contaminants from grains and cereals. Enhances food safety by reducing the levels of toxic substances in raw ingredients. Bentonite clay is used as an anti-caking agent in powdered and granular food products. And improves the handling and usability of powdered and granular foods by preventing moisture-induced clumping. It is also used as a stabilizer in various food formulations in certain processed foods and sauces to maintain consistency and prevent separation. Bentonite clay is occasionally used as a food additive, stabilizer, or thickener in certain formulations. It can also be used in health foods and supplements for its purported detoxification properties.

6.4.4 Activated Alumina

A highly porous form of aluminum oxide (Al_2O_3) is known for its excellent adsorption properties. It is used in the food industry for a variety of purposes due to its ability to adsorb water, contaminants, and impurities. Activated alumina is used to purify water by adsorbing contaminants and impurities. In the treatment of water used in food and beverage production, activated alumina removes fluoride, arsenic, and other contaminants. It is specifically used to remove fluoride from drinking water. In areas with high fluoride levels, activated alumina filters are used to reduce fluoride concentration from water supplied to food processing facilities and bottling plants. It is also used in the food industry to remove heavy metals such as lead and cadmium from water used in processing plants, and enhances the safety and purity of water and ingredients used in food production.

Activated alumina is used as a desiccant to control moisture levels in packaging and storage. And maintains the quality and shelf life of food products by controlling humidity and preventing spoilage. It serves as a catalyst or catalyst support in various chemical reactions in food processing and enhances the efficiency of chemical processes involved in food production. Activated alumina is used in air filtration systems to remove contaminants from the air in food processing and storage environments. Ensures a clean and safe processing environment, which is crucial for maintaining food quality and safety. For example, it is used to remove various impurities from food ingredients and additives, like the processing of edible oils and sweeteners. Activated alumina adsorbs colorants and contaminants to improve the purity and quality of processed food products by removing unwanted substances. It is also used in food packaging to control moisture and extend shelf life. Activated alumina can be integrated into packaging materials to absorb excess moisture and protect sensitive products to prevent spoilage, and maintain the freshness of packaged foods.

6.4.5 Silica Gel

A porous form of silicon dioxide (SiO_2) is commonly used in the food industry for its moisture-absorbing properties. This is used to control the moisture levels of packaged food products. Silica gel packets are placed in packaging for dried fruits, snacks, spices, and powdered foods to absorb excess moisture. Moreover, silica gel also helps to maintain the texture and quality of food products by controlling humidity.

Silica gel is used to protect food products during storage and transportation by controlling humidity levels. In packaging for odor-sensitive products, it acts as a natural odor absorber, ensuring that food items retain their desired flavors and remain free from unwanted smells. Silica gel as an adsorbent material can be utilized in storage bins or silos for grains, sugar, and other bulk dry goods to prevent moisture-related issues and to ensure that food products remain in good condition during long-term storage.

This is also used in certain drying processes to maintain the right humidity levels for getting optimal drying conditions, enhancing the efficiency and effectiveness of food processing operations. These adsorbent materials can be incorporated into packaging materials designed for regulated food products, as well as in packaging materials for pharmaceuticals and supplements that require precise moisture control.

6.5 CHALLENGES AND FUTURE PROSPECTS

6.5.1 Environmental Sustainability and Scalability

Despite the advancements in novel adsorbent materials, challenges remain in scaling up green synthesis methods for industrial applications. Research related to this needs to be focused on developing low-cost, scalable processes that could minimize environmental impact while maintaining high adsorption efficiency.

6.5.2 Functionalization and Target-Specific Adsorbents

The future of adsorbent technology lies in functionalization, where adsorbents are designed with specific chemical functionalities to target emerging contaminants.

This requires further research for the development of adsorbent materials that can selectively capture pharmaceuticals, pesticides, microplastics, and other emerging pollutants from complex matrices.

6.5.3 Regeneration and Reusability

One of the key challenges in adsorbent technology is the regeneration and reusability of adsorbents. Future research should focus on developing materials that can be easily regenerated with minimal energy consumption and without generating secondary pollutants.

6.6 CONCLUSION

The synthesis of advanced adsorbent materials using green and novel technologies holds immense promise for addressing environmental and industrial challenges. Green techniques such as biomass-derived adsorbents, hydrothermal carbonization, and plant-based synthesis offer sustainable alternatives to traditional adsorbent materials. Meanwhile, novel techniques such as MOFs, COFs, MIPs, and 3D printing have opened new avenues for creating highly efficient, selective, and customizable adsorbents for various applications. Continued research on environmental impact, scalability, and functionalization of these materials will be crucial in realizing their full potential in water purification, gas storage, and biomedical applications.

REFERENCES

Abdulfatai, J., Saka, A. A., Afolabi, A. S., & Micheal, O. (2013). Development of adsorbent from banana peel for wastewater treatment. *Applied Mechanics and Materials*, *248*, 310–315.

Aboagye, E. A., Santos, M. L. S., dos Santos Lins, P. V., Meili, L., Franco, D. S. P., Georgin, J., & Selvasembian, R. (2024). Green synthesized bio-nanomaterials for pollutant remediation. *Nanotechnology for Environmental Management*, *83*, 98.

Ahmed, M. B., Johir, M. A. H., Zhou, J. L., Ngo, H. H., Nghiem, L. D., Richardson, C., … & Bryant, M. R. (2019). Activated carbon preparation from biomass feedstock: Clean production and carbon dioxide adsorption. *Journal of Cleaner Production*, *225*, 405–413.

Akhtar, M. S., Ali, S., & Zaman, W. (2024). Innovative adsorbents for pollutant removal: Exploring the latest research and applications. *Molecules*, *29*(18), 4317.

Akinhanmi, T. F., Ofudje, E. A., Adeogun, A. I., Aina, P., & Joseph, I. M. (2020). Orange peel as low-cost adsorbent in the elimination of Cd (II) ion: Kinetics, isotherm, thermodynamic and optimization evaluations. *Bioresources and Bioprocessing*, *7*, 1–16.

Alatalo, S. M., & Sillanpää, M. (2020). Hydrothermal carbonization in the synthesis of sustainable porous carbon materials for water treatment. In *Advanced Water Treatment* (pp. 445–503). Elsevier.

Almeida, R. M., & Gonçalves, M. C. (2021). Sol–Gel process and products. *Encyclopedia of Glass Science, Technology, History, and Culture*, *2*, 969–979.

Al-Qodah, Z., & Al-Shannag, M. (2017). Heavy metal ions removal from wastewater using electrocoagulation processes: A comprehensive review. *Separation Science and Technology*, *52*, 2649–2676.

Al-Qodah, Z., Yahya, M.A., Al-Shannag, M. (2017). On the performance of bioadsorption processes for heavy metal ions removal by low-cost agricultural and natural by-products bioadsorbent: A review. *Desalination and Water Treatment*, *85*, 339–357.

Alshameri, A., He, H., Zhu, J., Xi, Y., Zhu, R., Ma, L., & Tao, Q. (2018). Adsorption of ammonium by different natural clay minerals: Characterization, kinetics and adsorption isotherms. *Applied Clay Science*, *159*, 83–93.

Antonopoulou, M., Tzamaria, A., Pedrosa, M. F., Ribeiro, A. R., Silva, A. M., Kaloudis, T., & Vlastos, D. (2024). Spirulina-based carbon materials as adsorbents for drinking water taste and odor control: Removal efficiency and assessment of cyto-genotoxic effects. *Science of the Total Environment*, *927*, 172227.

Avci, S. K., & Erucar, I. (2018). Porous materials. *Comprehensive Energy Systems*, *2*, 182–203.

Bachrun, S., AyuRizka, N., Annisa, S., & Arif, H. (2016). Preparation and characterization of activated carbon from sugarcane bagasse by physical activation with CO2 gas. In *IOP Conference Series: Materials Science and Engineering* (Vol. 105, No. 1, p. 012027). IOP Publishing.

Ballesteros-Gómez, A., Rubio, S., & Pérez-Bendito, D. (2009). Analytical methods for the determination of bisphenol A in food. *Journal of Chromatography A*, *1216*(3), 449–469.

Bruch, L. W., Cole, M. W., & Zaremba, E. (2007). *Physical Adsorption: Forces and Phenomena*. Courier Dover Publications.

Camargo, F. P., do Prado, P. F., Tonello, P. S., Dos Santos, A. C. A., & Duarte, I. C. S. (2018). Bioleaching of toxic metals from sewage sludge by co-inoculation of Acidithiobacillus and the biosurfactant-producing yeast Meyerozyma guilliermondii. *Journal of Environmental Management*, *211*, 28–35.

Cantarella, M., Carroccio, S. C., Dattilo, S., Avolio, R., Castaldo, R., Puglisi, C., & Privitera, V. (2019). Molecularly imprinted polymer for selective adsorption of diclofenac from contaminated water. *Chemical Engineering Journal*, *367*, 180–188.

Castro, L. E. N. D., Battocchio, D. A. J., Ribeiro, L. F., & Colpini, L. M. S. (2022). Development of adsorbent materials using residue from coffee industry and application in food dye adsorption processes. *Brazilian Archives of Biology and Technology*, *66*, e23210125.

Cheng, H., Zhu, Q., & Xing, Z. (2019). Adsorption of ammonia nitrogen in low temperature domestic wastewater by modification bentonite. *Journal of Cleaner Production*, *233*, 720–730.

Chintokoma, G. C., Chebude, Y., Kassahun, S. K., Demesa, A. G., & Koiranen, T. (2024). Sol-gel synthesis of composite adsorbent coating from Prosopis juliflora–activated carbon for simultaneous adsorptive removal of Cd2+ and Cr2O72-from wastewater. *AQUA—Water Infrastructure, Ecosystems and Society*, *73*(5), 945–968.

Chong, M. N., Jin, B., Chow, C. W.K., & Saint, C. (2010) Recent developments in photocatalytic water treatment technology: A review. *Water Research*, *4*, 2997–3027.

Devrajani, S. K., Ahmed, Z., Qambrani, N. A., Kanwal, S., Sundaram, U. M., & Mubarak, N. M. (2024). Mechanism of arsenic removal using brown seaweed derived impregnated with iron oxide biochar for batch and column studies. *Scientific Reports*, *14*(1), 18102.

Egbuchunam, T. O., Obi, G., Okieimen, F. E., & Yetgin, S. (2018). Low-cost materials for the removal of contaminants from wastewater. In *Modern Physical Chemistry: Engineering Models, Materials, and Methods With Applications* (pp. 434–448). Apple Academic Press.

Gao, Y., Yue, Q., Gao, B., & Li, A. (2020). Insight into activated carbon from different kinds of chemical activating agents: A review. *Science of the Total Environment*, *746*, 141094.

García-Mendieta, A., Olguín, M. T., & Solache-Ríos, M. (2012). Biosorption properties of green tomato husk (Physalis philadelphica Lam) for iron, manganese and iron–manganese from aqueous systems. *Desalination*, *284*, 167–174.

Gautam, S., & Saini, G. (2020). Use of natural coagulants for industrial wastewater treatment. *Global Journal of Environmental Science and Management*, *6*(4). 553–578.

Gebretsadik, H., Gebrekidan, A., & Demlie, L. (2020). Removal of heavy metals from aqueous solutions using Eucalyptus Camaldulensis: An alternate low-cost adsorbent. *Cogent Chemistry*, *6*(1), 1720892.

Gong, X. (2013). *Modification and Utilization of Sewage Sludge-Based Activated Carbon as Metal Adsorbents*. Doctoral dissertation (University of British Columbia).

Hernández-Montoya, V., García-Servin, J., & Bueno-López, J. I. (2012). Thermal treatments and activation procedures used in the preparation of activated carbons. In *Lignocellulosic Precursors used in the Synthesis of Activated Carbon, Characterization Techniques and Applications in the Wastewater Treatment* (pp. 19–36). IntechOpen.

Ibrahim, W. M., Hassan, A. F., & Azab, Y. A. (2016). Biosorption of toxic heavy metals from aqueous solution by Ulva lactuca activated carbon. *Egyptian Journal of Basic and Applied Sciences*, *3*(3), 241–249.

Iwuozor, K. O. (2019). Prospects and challenges of using coagulation-flocculation method in the treatment of effluents. *Advanced Journal of Chemistry-Section A*, *2*(2), 105–127.

Jin, H., Zhang, Y., Zhang, X., Chang, M., Li, C., Lu, X., & Wang, Q. (2022). 3D printed geopolymer adsorption sieve for removal of methylene blue and adsorption mechanism. *Colloids and Surfaces A: Physicochemical and Engineering Aspects*, *648*, 129235.

Li, H., & Yang, Z. (2015). Metalorganic frameworks as efficient adsorbents for environmental pollutants. *Environmental Science & Technology*, *49*(6), 36063618.

Li, M., & Xiao, R. (2019). Preparation of a dual pore structure activated carbon from rice husk char as an adsorbent for CO2 capture. *Fuel Processing Technology*, *186*, 35–39

Li, S., De Silva, T., Arsano, I., Gallaba, D., Karunanithy, R., Wasala, M., … & Talapatra, S. (2020). High adsorption of benzoic acid on single walled carbon nanotube bundles. *Scientific Reports*, *10*(1), 10013.

Lin, R. B., Xiang, S., Zhou, W., & Chen, B. (2020). Microporous metal-organic framework materials for gas separation. *Chem*, *6*(2), 337–363.

Liu, Y., Yan, C., Zhao, J., Zhang, Z., Wang, H., Zhou, S., & Wu, L. (2018). Synthesis of zeolite P1 from fly ash under solvent-free conditions for ammonium removal from water. *Journal of Cleaner Production*, *202*, 11–22.

Mai, N. T., Nguyen, M. N., Tsubota, T., Nguyen, P. L., & Nguyen, N. H. (2021). Evolution of physico-chemical properties of Dicranopteris linearis-derived activated carbon under various physical activation atmospheres. *Scientific Reports*, *11*(1), 14430.

Mochizuki, T., Kubota, M., Matsuda, H., & Camacho, L. F. E. (2016). Adsorption behaviors of ammonia and hydrogen sulfide on activated carbon prepared from petroleum coke by KOH chemical activation. *Fuel Processing Technology*, *144*, 164–169.

Owsianiak, M., Ryberg, M. W., Renz, M., Hitzl, M., & Hauschild, M. Z. (2016). Environmental performance of hydrothermal carbonization of four wet biomass waste streams at industry-relevant scales. *ACS Sustainable Chemistry & Engineering*, *4*(12), 6783–6791.

Perpetuo, E. A., Souza, C. B., & Nascimento, C. A. O. (2011). Engineering bacteria for bioremediation. *Progress in Molecular and Environmental Bioengineering*, 20, 605–632.

Prathna, T. C. (2019). Biosynthesized nanoparticles for water treatment. In *Biological Synthesis of Nanoparticles and Their Applications* (pp. 227–232). CRC Press.

Pu, S., Xue, S., Yang, Z., Hou, Y., Zhu, R., & Chu, W. (2018). In situ co-precipitation preparation of a superparamagnetic graphene oxide/Fe3O4 nanocomposite as an adsorbent for wastewater purification: Synthesis, characterization, kinetics, and isotherm studies. *Environmental Science and Pollution Research*, *25*, 17310–17320.

Rashed, M. N., Soltan, M. E., Ahmed, M. M., & Abdou, A. (2018). Heavy metals removal from wastewater by adsorption on modified physically activated sewage sludge. *Archives of Organic and Inorganic Chemical Sciences*, *1*, 18–25.

Rouquerol, J., Rouquerol, F., Llewellyn, P., Maurin, G., & Sing, K. (2013). *Adsorption by Powders and Porous Solids: Principles, Methodology and Applications*. Academic Press.

Saleh, A. M., Mahdi, H. H., Alias, A. B., Abd Hadi, N. K., Qarizada, D., Jawad, A. H., & Saleh, N. M. (2024). Equilibrium and kinetic studies in adsorption of H2S using coconut shell activated carbon xerogel: Effect of mass adsorbent and temperature. *Desalination and Water Treatment*, *317*, 100149.

Salman, S. D., Rasheed, I. M., & Mohammed, A. K. (2021). Adsorption of heavy metal ions using activated carbon derived from Eichhornia (water hyacinth). In *IOP Conference Series: Earth and Environmental Science* (Vol. 779, No. 1, p. 012074). IOP Publishing.

Sanchez, R. B. M. (2014). *Synthesis and Characterization of Oxide and Nickel Metal Nanocomposites for Use as Catalysts for Carbon.* Nanotubes Thesis (Postgraduate Program in Materials and Chemical and Metallurgical Process Engineering) Pontifical Catholic University of Rio de Janeiro, Rio de Janeiro - RJ (in Portuguese).

Serrano-Gómez, J., López-González, H., Olguín, M. T., & Bulbulian, S. (2015). Carbonaceous material obtained from exhausted coffee by an aqueous solution combustion process and used for cobalt (II) and cadmium (II) sorption. *Journal of Environmental Management, 156*, 121–127.

Sharma, P., Prakash, J., & Kaushal, R. (2022). An insight into the green synthesis of SiO2 nanostructures as a novel adsorbent for removal of toxic water pollutants. *Environmental Research, 212*, 113328.

Shen, Z., Zhang, Z., Zhang, M., Rinklebe, J., Ma, Y., & Hou, D. (2021). Effect of production temperature and particle size of rice husk biochar on mercury immobilization and erosion prevention of a mercury contaminated soil. *Journal of Hazardous Materials, 420*, 126646.

Singh, S., Khan, N. A., Shehata, N., Singh, J., & Ramamurthy, P. C. (2024). Insight into biochar as sustainable biomass: Production methods, characteristics, and environmental remediation. *Journal of Cleaner Production*, 475, 143645.

Syam Babu, D., Anantha Singh, T. S., Nidheesh, P. V., & Suresh Kumar, M. (2020). Industrial wastewater treatment by electrocoagulation process. *Separation Science and Technology, 55*(17), 3195–3227.

Tawfik, S. (2022). Adsorption technology and surface science. *Interface Science and Technology, 34*, 39–64.

Tran, H. N., Chao, H. P., & You, S. J. (2018). Activated carbons from golden shower upon different chemical activation methods: Synthesis and characterizations. *Adsorption Science & Technology, 36*(1–2), 95–113.

Tsai, W. T., & Jiang, T. J. (2018). Mesoporous activated carbon produced from coconut shell using a single-step physical activation process. *Biomass Conversion and Biorefinery, 8*, 711–718.

Upadhyay, M., Hansa, A., Devi, A., Lajayer, B. A., Price, G. W., & Sharma, R. (2024). Status and future scope of algal biomass-based remediation for various environmental contaminants. *Journal of Water Process Engineering, 65*, 105809.

Vieira, A. M. S., Vieira, M. F., Silva, G. F., Araújo, Á. A., Fagundes-Klen, M. R., Veit, M. T., & Bergamasco, R. (2010). Use of Moringa oleifera seed as a natural adsorbent for wastewater treatment. *Water, Air, and Soil Pollution, 206*, 273–281.

Wang, B., Lan, J., Bo, C., Gong, B., & Ou, J. (2023). Adsorption of heavy metal onto biomass-derived activated carbon. *RSC Advances, 13*(7), 4275–4302.

Wang, X. Y., Li, D. X., Yang, B. M., & Li, W. (2013). Textural characteristics of coconut shell-based activated carbons with steam activation. *Advanced Materials Research, 608*, 366–373.

Wei, Y., Wei, S., Liu, C., Chen, T., Tang, Y., Ma, J., ... & Luo, S. (2019). Efficient removal of arsenic from groundwater using iron oxide nanoneedle array-decorated biochar fibers with high Fe utilization and fast adsorption kinetics. *Water Research, 167*, 115107.

Wu, Q., Zhang, Y., Cui, M. H., Liu, H., Liu, H., Zheng, Z., ... & Wen, D. (2022). Pyrolyzing pharmaceutical sludge to biochar as an efficient adsorbent for deep removal of fluoroquinolone antibiotics from pharmaceutical wastewater: Performance and mechanism. *Journal of Hazardous Materials, 426*, 127798.

Zhao, Y., & Guo, X. (2012). Graphene oxide and its derivatives for adsorption of organic molecules and ions: A review. *Carbon, 50*(14), 46174633.

7 Advanced Oxidation Processes for Wastewater Treatment

Amit Kumar Behera and Krushna Prasad Shadangi

7.1 INTRODUCTION

Water pollution is an emerging concern because of the release of various recalcitrant pollutants into water bodies. Among the various pollutants, recalcitrant organic compounds cause major adverse effects on aquatic and human lives. Recalcitrant compounds like organic dyes, acetaminophen, and phenol are released into water bodies as effluents by paper, textile, pharmaceutical, paint, and leather industries (Yang et al., 2022). The recalcitrant compounds released are stable, carcinogenic, and non-biodegradable (Wan and Wang, 2021). Among the dyes released from industries, Rhodamine B dye and phenol have a maximum tolerable limit of 10 ng/L for safe use. Rhodamine B dye resists biodegradation, which results in a longer presence of dye in water. Phenol is carcinogenic, which disrupts the endocrine system and disorders nervous system (Kasonga et al., 2021). The presence of pharmaceuticals like acetaminophen in the environment, especially in surface waters, is becoming an increasingly pressing issue. Acetaminophen enters the environment primarily through human excretion of unmetabolized or partially metabolized drugs, improper disposal of expired medications, and the use of antibiotics in livestock farming. Acetaminophen is also harmful to aquatic life, such as fish and invertebrates, and contributes to the development of antibiotic-resistant bacteria. Wastewater treatment plants, which traditionally use activated sludge, ineffectively remove acetaminophen because they were not designed to process such foreign compounds (Li et al., 2021). Coagulation, filtration, adsorption, and membrane separation are the traditional treatment methods used for treating recalcitrant compounds like Rhodamine b, acetaminophen, and phenol. However, the cost involved in the treatment processes and the generation of secondary pollutants are the limiting factors. Advanced oxidation processes (AOPs) are an emerging technology for the complete degradation of Rhodamine B, acetaminophen, and phenol (Moscow et al., 2022). AOPs are energy-efficient wastewater treatment methods for the complete remediation of organic recalcitrant compounds.

 DOI: 10.1201/9781003767251-7

AOPs are one of the most promising and effective methods for the elimination of organic contaminants from wastewater. Industrial discharges into the environment increasingly show a rising number of hazardous substances. These contaminants frequently cause major environmental and biological harm and pose major risks to aquatic ecosystems and human health. Thus, putting effective treatment plans into practice is essential. AOPs are a good way to remediate wastewater that contains these dangerous substances. In order to break down persistent organic pollutants, these technologies rely on the in-situ creation of highly reactive oxidizing species, such as hydroxyl and sulfate radicals (Jayakumar et al., 2022). Some AOP systems add ozone and ultraviolet (UV) radiation to improve therapeutic efficacy. People have widely implemented AOPs to purify drinking water. Moreover, a variety of AOP techniques—like electrochemical methods, electron beam technology, plasma, microwave (MW), and ultrasound-assisted processes—are actively being explored by researchers in environmental science.

AOPs follow a series of fundamental steps:

- Initially, the reaction generates powerful oxidizing agents such as hydroxyl radicals (•OH), hydroperoxyl radicals (HO_2•), and superoxide anions ($O_2^{\bullet -}$).
- These reactive species then interact with the organic pollutants in the wastewater, converting them into simpler, less hazardous, and biodegradable compounds.
- In the final stage, these intermediate compounds are further oxidized, resulting in their complete mineralization of water, carbon dioxide, and inorganic salts.

AOPs are particularly valuable because they help bridge the performance gap between conventional physicochemical or biological treatments and the increasingly stringent environmental regulations (Akash et al., 2022). Although research into AOPs has been extensive over the past two decades, these technologies are now considered well-developed. Nevertheless, certain aspects still offer potential for further investigation and innovation.

7.2 TYPES OF ADVANCED OXIDATION PROCESSES

Researchers use many methods to classify AOPs as non-photochemical and photochemical. The different non-photochemical and photochemical AOPs are shown in Figure 7.1. Photochemical AOPs are widely employed as effective methods for degrading organic contaminants in wastewater. The synergistic action of UV light and hydrogen peroxide significantly enhances the elimination of a diverse array of pollutants (Amin et al., 2014). Commonly applied photochemical AOPs include UV irradiation alone, UV combined with H_2O_2, UV-Fe^{2+}, UV-H_2O_2-Fe^{2+}, UV-O_3, UV-$S_2O_8^{2-}$, UV-TiO_2, UV-chlorine, and other UV-based systems using various photocatalysts (Patil and Gogate, 2015). These UV-driven processes can generate both hydroxyl and sulfate radicals, leading to improved removal efficiencies in a broad spectrum of contaminants.

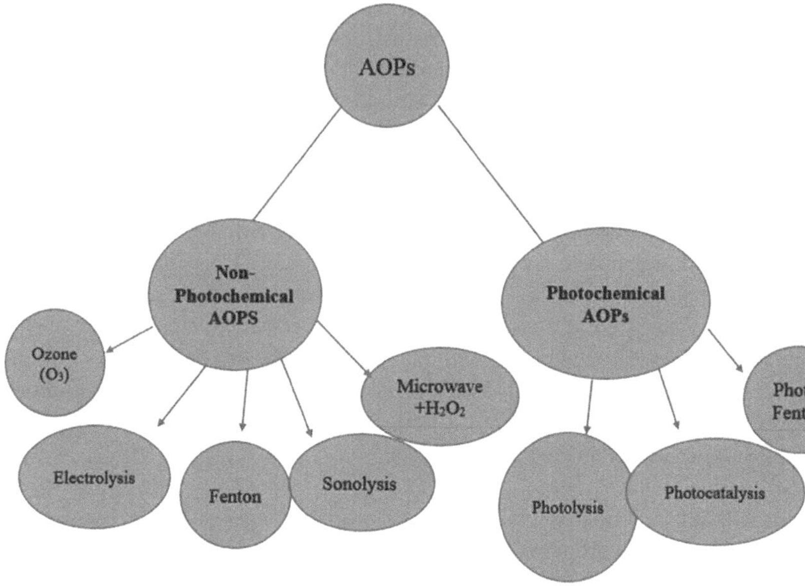

FIGURE 7.1 Types of advanced oxidation processes.

7.2.1 Non-Photochemical AOPs

7.2.1.1 Fenton Oxidation

One of the most sophisticated methods for treating wastewater is Fenton oxidation (Figure 7.2). This method depends on the reaction of hydrogen peroxide with ferrous ions, which produces hydroxyl radicals (•OH). Because of their high reactivity, these radicals can decompose hazardous and tenacious organic materials in wastewater. Fenton disclosed that ferrous salts could activate H_2O_2 to oxidize tartaric acid in 1894; he first identified the Fenton reaction (Boczkaj and Fernandes, 2017). However, it was not until the late 1960s that it was discovered as a way to break down harmful organic contaminants. Fe^{2+} ions catalyze H_2O_2 in Fenton treatment, producing hydroxyl radicals and other reactive species that collectively aid in the full oxidation of organic contaminants. The hydroxyl radical possesses a high oxidative potential (standard reduction potential of 2.80 V relative to the standard hydrogen electrode [SHE]), enabling it to degrade resistant organic contaminants in wastewater, often fully mineralizing them into carbon dioxide, water, and inorganic salts. H_2O_2 oxidizes Fe (II) to Fe (III) during the Fenton reaction, producing a hydroxyl radical and a hydroxyl anion.

The peroxide radical and hydrogen peroxide proton (iron III) is subsequently reduced back to iron (II) (disproportionation). Fenton's reagent produces a hydroxyl free radical, which has a high oxidation potential (2.80 V) and functions as a non-selective oxidant (Equations 7.1 and 7.2) (Novotna et al., 2019). Fenton's reagent oxidizes organic compounds quickly and exothermically.

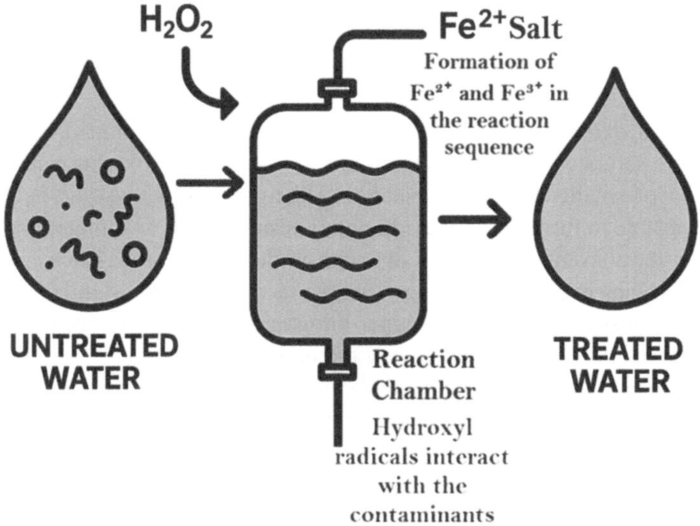

FIGURE 7.2 Fenton oxidation process.

$$Fe^{+2} + H_2O_2 \rightarrow Fe^{+3} + OH^{\cdot} + OH^{-} \quad (7.1)$$

$$Fe^{+3} + H_2O_2 \rightarrow Fe^{+2} + OOH^{\cdot} + H^{+} \quad (7.2)$$

One of the known metal-catalyzed oxidation processes of organic molecules that are water soluble is Fenton's oxidation. It quickly oxidized organic substances. By forming ferric-hydroxo complexes, the Fenton's reagent also exhibits coagulation properties. Therefore, by improving the removal effectiveness of Fenton's oxidation, the coagulation phase serves as a polishing step. The following is a representation of the Fenton's oxidation reaction (Parsons, 2004):

$$2Fe^{+2} + H_2O_2 + 2H^{+} \rightarrow 2Fe^{+3} + 2H_2O \quad (7.3)$$

These equations state that to produce the most hydroxyl radicals necessary to oxidize the organic molecules, the pH of the solution should be in the acidic range. The production of hydroxyl radicals gets slowed down at higher pH values (>4) because of the formation of ferric-hydroxo complexes. Since H_2O_2 is unstable in alkaline solutions, its oxidation property is limited by pH values over 7 (Ameta et al., 2013).

7.2.1.2 Ozone Oxidation

Great energy consumption and limited electrode lifespan also hinder the long-term use of some AOPs. Among these, ozonation has attracted significant attention in recent years due to the strong oxidizing power of ozone (O_3), with a standard oxidation potential of 2.07 V versus the SHE (Yahya et al., 2014). It has been widely utilized in drinking water decontamination and wastewater management. Ozone can directly reduce organic compounds through mechanisms such as electrophilic and nucleophilic reactions, as well as cycloaddition (Figure 7.3).

The direct oxidation capacity of ozone (O_3) is relatively limited when dealing with certain saturated and polycyclic hydrocarbons. As a result, this can lead to the generation of smaller byproducts such as aldehydes and carboxylic acids, which ozone is unable to further degrade, ultimately resulting in a low mineralization rate. Moreover, the conventional method of ozone production—corona discharge—exhibits low efficiency in converting oxygen into ozone and demands significant electrical energy, consuming roughly 15 kWh per kilogram of ozone generated (Noor et al., 2022). The combination of high energy requirements and ozone's low solubility in water significantly hinders its wider application in treatment processes. Nonetheless, ozone can also initiate chain reactions that produce hydroxyl radicals (•OH) indirectly. Due to their high oxidation potential (2.80 V vs. SHE), non-selectivity, and rapid reactivity, these radicals can completely oxidize many persistent organic pollutants (Pandian et al., 2021). Sludge reduction and the elimination of stubborn organic pollutants from wastewater are two advantages of ozonation in wastewater treatment facilities. Ozonation has been shown to cause sludge solubilization, which lowers the total biomass yield. Numerous elements, which can be categorized into hydrodynamic and physicochemical impacts, affect the mass transport of ozone. Due to its instability and rapid breakdown into molecular oxygen, ozone has a poor usage rate. Furthermore, ozone has a poor reaction rate and does not completely oxidize certain refractory organic molecules. To boost the formation of hydroxyl radicals and hence the treatment efficacy, ozone is mixed with H_2O_2, UV radiation, catalysts, photocatalysts, and ultrasound (Khairnar et al., 2013).

7.2.1.3 Electrolysis-Based AOPs

Electrochemical advanced oxidation processes (EAOPs), as shown in Figure 7.4, have garnered growing interest as a promising subset of AOPs. Among these, anodic oxidation (AO) is the most straightforward and widely used. In AO, organic contaminants can be oxidized directly at the anode surface through electron transfer or indirectly by hydroxyl radicals (•OH) that are weakly adsorbed on the anode, as well as by reactive species present in the bulk solution—such as active chlorine, ozone (O_3), persulfates, and hydrogen peroxide (H_2O_2). When AO is coupled with the cathodic generation of H_2O_2, the method is referred to as AO with electro-generated

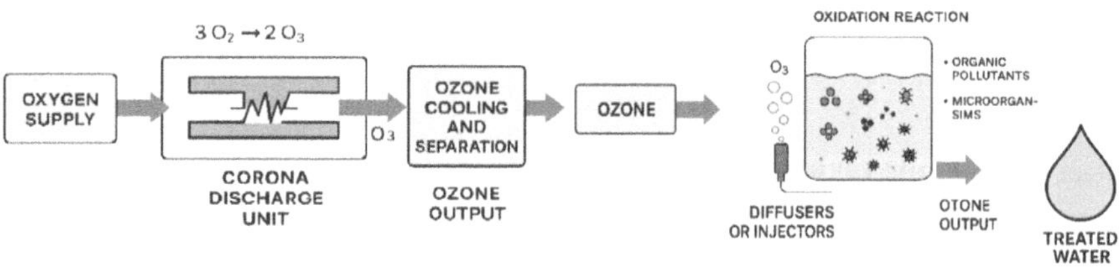

FIGURE 7.3 Ozone oxidation process for wastewater treatment.

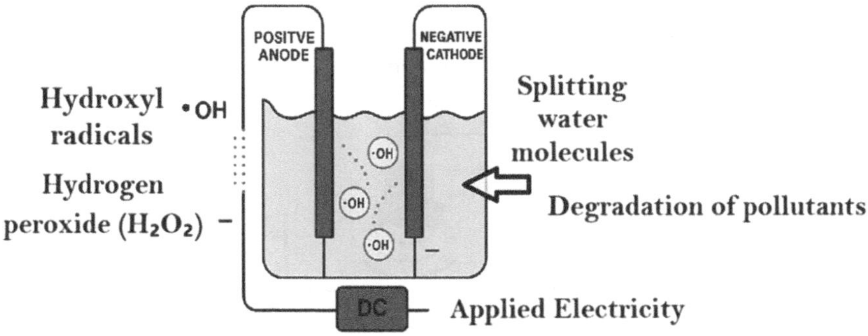

FIGURE 7.4 Electrolysis-based advanced oxidation process for wastewater treatment.

H_2O_2 (AO-H_2O_2) (Ali et al., 2013). Introducing Fe^{2+} into the system enables the formation of hydroxyl radicals via the Fenton reaction, forming the basis of the widely studied electro-Fenton (EF) process.

The combination of electrolysis with other AOP variants like photoelectron-Fenton and solar photoelectron-Fenton, which integrate EF with artificial or solar irradiation, is of major interest nowadays. Apart from these, other EAOPs, such as peroxi-coagulation, Fered-Fenton, electrochemical peroxidation, and sonoelectro-Fenton, have been used for treating various types of wastewater. Given the high capital investment and operational costs associated with EAOPs, there has been an emphasis on combining them with other treatment technologies—such as biological treatments, chemical coagulation, electrocoagulation, and membrane-based methods—to enhance cost-effectiveness and overall treatment performance (Benjamin et al., 2011).

In this process, the water flows through an electrolytic cell made up of titanium or graphite electrode, and by the application of electricity, the splitting of water molecules generates reactive species such as Hydroxyl radicals (•OH), oxygen, and hydrogen peroxide (H_2O_2), which react with the organic pollutants such as pesticides, pharmaceuticals, and dyes and produce CO_2 and salts. The advantage of this process is that it does not need any external chemicals and also does not generate any secondary pollutants. However, it is an energy-intensive process and needs the replacement of electrodes after a period of operation.

7.2.1.4 Sonolysis

Sonolysis is a relatively recent technique within AOPs, which utilizes ultrasound irradiation to generate hydroxyl radicals (•OH) through the pyrolytic breakdown of water. It is often regarded as an environmentally friendly or "green" method for removing toxic pollutants from water, as it does not require or introduce any external chemicals into the water system. However, the process demands significant energy input to create acoustic cavitation bubbles induced by ultrasound waves (Figure 7.5). The formation and behavior of these bubbles are strongly influenced by the ultrasound parameters, particularly the power and frequency, which directly affect the degradation efficiency of contaminants (Bhosale et al., 2013). One of the

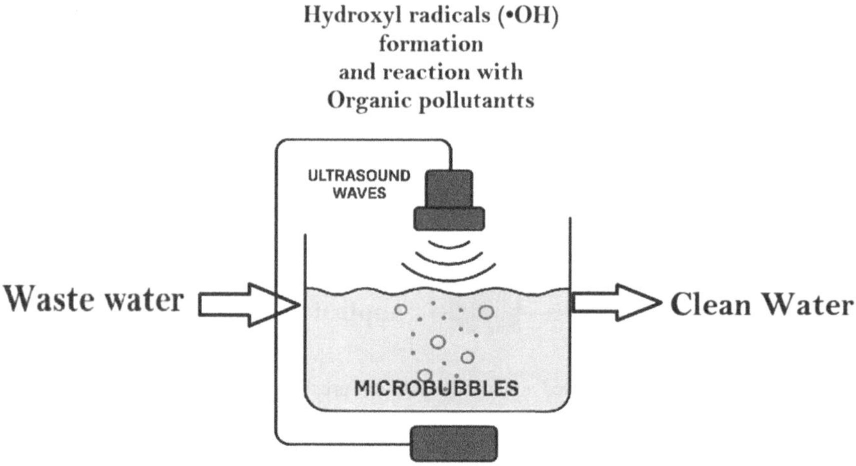

FIGURE 7.5 Ultrasound-based wastewater treatment process.

key advantages of sonolysis is that it initially operates without the need for external chemicals to produce hydroxyl radicals. However, achieving effective mineralization of pollutants often requires high energy input and the addition of chemical agents. As a result, its application is generally limited to laboratory-scale studies or the treatment of wastewater with low initial chemical oxygen demand (COD). To overcome the limitations related to incomplete mineralization, combining sonolysis with other AOPs is often suggested to improve overall treatment outcomes and reduce COD levels in the final effluent.

Chatterjee and Dasgupta (2005) found that increasing power density (from 100 to 400 W/L) enhanced the degradation of diclofenac and carbamazepine, suggesting that power density may play a more significant role than the frequency in some cases.

Higher ultrasonic power results in a linear increase in pollutant degradation, attributed to the formation of more cavitation bubbles and a higher yield of •OH radicals. The addition of radical promoters (Sun et al., 2020) or other compounds such as mannitol and calcium carbonate for enhancing antimicrobial activity and degrading oxacillin has also been explored.

7.2.1.5 Microwave +H_2O_2

MWs consist of perpendicular electric and magnetic fields that interact with materials, generating dielectric and magnetic losses that result in heat production (Figure 7.6). Hydroxyl radical-based advanced oxidation processes are well-established and highly effective methods for pollutant degradation. Hydrogen peroxide, a commonly used and clean oxidant in water treatment (Kumar and Chandra, 2020), is a typical source of hydroxyl radicals (OH•). Under MW irradiation, H_2O_2 can decompose to form OH• (Equation 7.4). The MW-H_2O_2 system has proven to be an efficient approach for removing environmental contaminants.

$$H_2O_2 \rightarrow 2OH\bullet(\text{Microwave assisted}) \tag{7.4}$$

MW radiation stimulates the electronic vibrations of molecules, raising the system's temperature in a shorter time and activating the production of more free radicals (Saharan et al., 2014). System temperature in the proper range for cooling equipment such as condensers, are installed within the MW reactor in practice. In general, the MW-H_2O_2 system shows the optimum oxidation efficiency under acidic or neutral conditions, while the MW-H_2O_2 system presents a scavenging effect of free radicals under alkaline conditions (Sharma and Philip, 2016). It was observed that landfill leachate was optimally treated at pH = 5, with COD removal of 47.37% (Lu et al., 2019).

7.2.2 Photochemical AOPs (Figure 7.6)

7.2.3 Photolysis

Photolysis—also known as photo dissociation or photodecomposition—is a chemical reaction where a molecule, whether organic or inorganic, is broken down by the energy of photons (Figure 7.7). This process involves the interaction of one or more photons with a target molecule. Photolysis is not limited to visible light; any photon with adequate energy, such as those from UV light, X-rays, or gamma rays, can initiate this reaction, as the energy of a photon is inversely related to its wavelength (Xia et al., 2017). It is important to distinguish photolysis from photosynthesis—the latter

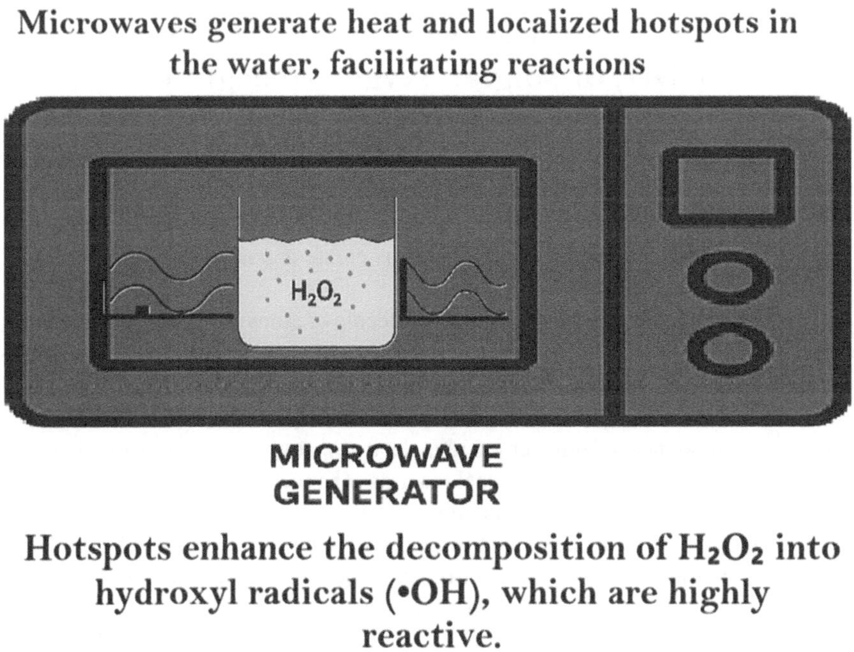

FIGURE 7.6 Microwave-assisted water treatment process.

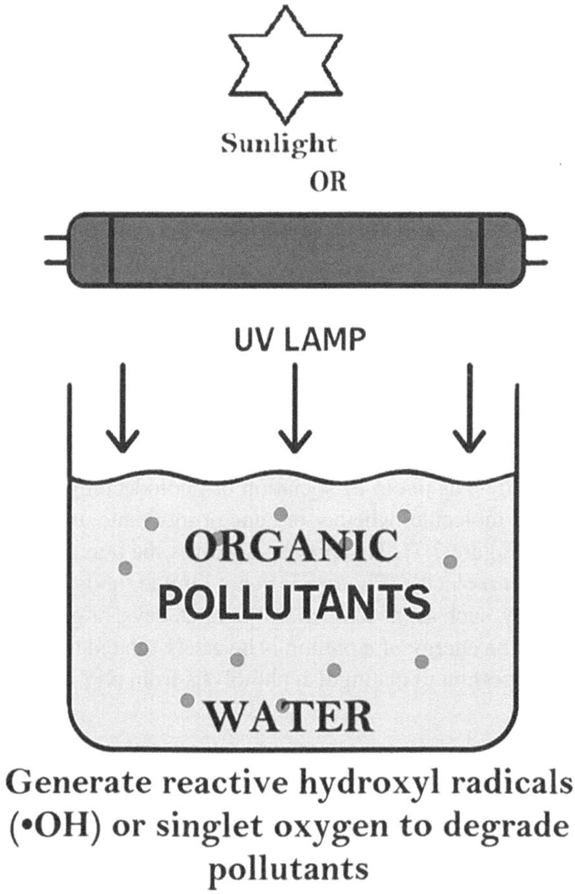

FIGURE 7.7 Photolysis process of water treatment.

being a biological process in which organisms create organic compounds as part of their life functions.

To initiate a photochemical reaction, a molecule or atom must absorb light energy carried by a photon. This absorption promotes the molecule into a temporarily excited state, altering its physical and chemical characteristics compared to its original form. Once in this excited state, the molecule may undergo structural changes, form new compounds, or transfer energy, electrons, atoms, or protons to other molecules—triggering a chain of chemical reactions (Kim et al., 1985).

When a molecule absorbs this energy, its energy level rises beyond the normal, marking the transition to an excited or activated state. For example, if a molecule absorbs visible or UV light, an electron in a low-energy state jumps to a higher energy level. In contrast, absorption of infrared radiation affects the vibrational motions of the nuclei within the molecule. Following this energy absorption, the molecule may undergo various initial (primary) photochemical reactions, which can be followed by additional (secondary) reactions, further extending the chemical transformation.

7.2.4 Photocatalysis

The words "photo" (from the word "photon") and "catalyst," which refers to a material that changes the rate of a reaction when it is present, are combined to form the term "photocatalyst." Consequently, substances that alter a chemical reaction's pace upon exposure to light are known as photocatalysts. Photocatalysis is the term for this phenomenon. Reactions that use a semiconductor and light are referred to as photocatalysis. A photocatalyst is a substrate that both absorbs light and catalyzes chemical reactions. In essence, all photocatalysts are semiconductors. A phenomenon known as photocatalysis occurs when a semiconducting substance is exposed to light, creating an electron–hole pair (Bhattacharjee and Ahmaruzzaman, 2015).

Based on how the reactants appear in their physical states, photocatalytic reactions can be divided into two categories. Homogeneous photocatalysis occurs when the reactant and semiconductor are in the same phase. Such photocatalytic reactions are referred to as homogeneous photocatalysis, regardless of the phase—gas, solid, or liquid. The other is known as the heterogeneous photocatalysis process, where the reactant and semiconductor are in distinct phases.

The band gap is the energy differential between the valence band (HOMO) and the conduction band (LUMO) (Kaushik et al., 2018). Therefore, the different materials are classified into conductors (band gap < 1.0 eV), insulators (band gap > 5.0 eV), and semi-conductors (band gap < 1.5–3.0 eV) (Mathew et al., 2014).

Heterogeneous catalysts are generally classified into two main types: unsupported (or bulk) catalysts and supported catalysts. Bulk catalysts include metal oxides such as Al_2O_3, SiO_2, and B_2O_3, as well as zeolite molecular sieves like ZSM-5 and various forms of zeolite, including types X, Y, and B. In contrast, supported catalysts utilize porous materials as a base, onto which active substances—typically metals or metal oxides—are deposited to form the catalytic surface. These catalysts play a crucial role in enhancing the rate of numerous chemical reactions, including carbon monoxide oxidation, selective oxidation, water-gas shift reactions, selective hydrogenation, electrocatalysis, photocatalysis, organic synthesis, nitrogen oxide removal (deNOx), and reforming processes (Zhang et al., 2020).

7.3 ADVANTAGES

AOPs offer several significant advantages compared to conventional wastewater treatment methods:

- They exhibit rapid reaction rates.
- They are capable of reducing toxicity and fully mineralizing organic pollutants.
- Unlike membrane filtration or activated carbon adsorption, AOPs do not concentrate contaminants for later disposal.

- They do not generate sludge, which is commonly produced in physical, chemical, or biological treatment systems.
- Their non-selective mechanism allows for the simultaneous degradation of various organic compounds.
- While the initial installation may be capital-intensive, operational costs are relatively low, and the process chemistry can be customized for specific treatment needs.

7.4 FUTURE OUTLOOKS

- AOPs driven by solar energy offer an important advantage in effectively treating industrial wastewater and simultaneously generating energy. This makes solar-driven AOPs a more cost-effective solution for water purification and reuse.
- Using industrial waste materials as catalysts—such as red mud, a byproduct from bauxite processing—can help reduce treatment costs. This approach, known as waste valorization, is a promising and underutilized concept that adds value to waste while supporting sustainable treatment solutions.
- Continuous innovation has led to the discovery of new materials with enhanced properties. As a result, AOPs have become a multidisciplinary topic involving engineering, science, and environmental studies. Collaboration among researchers from various fields is essential to effectively address water contamination challenges. Therefore, from an economic perspective, it is not practical to entirely replace conventional wastewater treatment methods with AOPs (Al Mayyahi and Al-Asadi, 2018). However, this challenge can be addressed by integrating AOPs with traditional treatment techniques. Such a combined approach is both feasible and more cost-effective compared to relying solely on AOPs for the entire treatment process.

REFERENCES

Akash, S., Sivaprakash, B., Rajamohan, N., Govarthanan, M., Elakiya, B.T., 2022. Remediation of pharmaceutical pollutants using graphene-based materials-A review on operating conditions, mechanism and toxicology. *Chemosphere* 306, 135520. https://doi.org/10.1016/j.chemosphere.2022.135520

Al Mayyahi, A., Al-Asadi, H.A.A., 2018. Advanced oxidation processes (AOPs) for wastewater treatment and reuse: A brief review. *Asian J. Appl. Sci. Technol.* 2, 18–30. https://doi.org/10.1016/j.cej.2009.09.042

Ali, M.A., Idris, M.R., Quayum, M.E., 2013. Fabrication of ZnO nanoparticles by solution combustion method for the photocatalytic degradation of organic dye. *J. Nanostruct. Chem.* http://doi.org/10.1186/2193-8865-3-36

Ameta, R., Kumar, A., Punjabi, P.B., Ameta, S.C., 2013. *Advanced Oxidation Processes: Basics and Applications in Waste Water Treatment: Advanced Processes and Technologies*, edited by Rao, D.G., Senthilkumar, R., Byrne, J.A., Feroz, S. (Eds.). IWA & CRC Press.

Amin, M.T., Alazba, A.A., Manzoor, U., 2014. A review of removal of pollutants from water/wastewater using different types of nanomaterials. *Adv. Mater. Sci. Eng.* http://doi.org/10.1155/2014/825910

Benjamin, S., Vaya, D., Punjabi, P.B., Ameta, S.C., 2011. Enhancing photocatalytic activity of zinc oxide by coating with some natural pigments. *Arab. J. Chem.* 4, 205–209.

Bhattacharjee, A., Ahmaruzzaman, M., 2015. Facile synthesis of SnO_2 quantum dots and its photocatalytic activity in the degradation of eosin Y dye: A green approach. *Mater. Lett.* 139, 418–421.

Bhosale, R., Pujari, S., Muley, G., Pagare, B., Gambhire, A., 2013. Visible-light-activated nanocomposite photocatalyst of Cr_2O_3/SnO_2. *J. Nanostruct. Chem.* http://doi.org/10.1186/2193-8865-3-46

Boczkaj, G., Fernandes, A., 2017. Wastewater treatment by means of advanced oxidation processes at basic pH conditions: A review. *Chem. Eng. J.* 320, 608–633.

Chatterjee, D., Dasgupta, S., 2005. Visible light induced photocatalytic degradation of organic pollutants. *J. Photochem. Photobiol. C Rev.* 6 (2–3), 186–205.

Jayakumar, V., Govindaradjane, S., Rajamohan, N., Rajasimman, M., 2022. Biosorption potential of brown algae, Sargassum polycystum, for the removal of toxic metals, cadmium and zinc. *Environ. Sci. Pollut. Control Ser.* 29 (28), 41909–41922. https://doi.org/10.1007/s11356-021-15185-7

Kasonga, T.K., Coetzee, M.A., Kamika, I., Ngole-Jeme, V.M., Momba, M.N.B., 2021. Endocrine-disruptive chemicals as contaminants of emerging concern in wastewater and surface water: A review. *J. Environ. Manag.* 277, 111485. https://doi.org/10.1016/j.jenvman.2020.111485

Kaushik, A., Basu, S., Raturi, S., Batra, V.S., Balakrishnan, M., 2018. Recovery of antioxidants from sugarcane molasses distillery wastewater and its effect on biomethanation. *J. Water Proc. Eng.* 25, 205–211.

Khairnar, P., Chavan, F., Diware, V.R., 2013. Generation of energy from distillery waste water. *Int. J. Sci. Spiritual Bus Technol.* 2, 30–35.

Kim, S.B., Hayase, F., Kato, H., 1985. Decolourisation and degradation products of melanoidin on ozonolysis. *Agric. Biol. Chem.* 49, 785–792.

Kumar, V., Chandra, R., 2020. Bioremediation of melanoidins containing distillery waste for environmental safety. In: Saxena, G., Bharagava, R.N. (Eds.), *Bioremediation of Industrial Waste for Environmental Safety. Vol II—Microbes and Methods for Industrial Waste Management*. Springer.

Li, Y., Dong, H., Li, L., Tang, L., Tian, R., Li, R., Chen, J., Xie, Q., Jin, Z., Xiao, J., Xiao, S., 2021. Recent advances in waste water treatment through transition metal sulfides-based advanced oxidation processes. *Water Res.* 192, 116850. https://doi.org/10.1016/j.watres.2021.116850

Lu, P., Cai, F., Zhang, J., Liu, Y., Sun, Y., 2019. Hydrogen production from methanol steam reforming over B-modified CuZnAlOx catalysts. *J. Fuel Chem. Technol.* 47, 791–798.

Mathew, S., Yella, A., Gao, P., Humphry Baker, R., Curchod, B.F.E., Astani, N.A., et al., 2014. Dye-sensitized solar cells with 13% efficiency achieved through the molecular engineering of porphyrin sensitizers. *Nature Chem.* 6(3), 242–247.

Moscow, S., Kavinkumar, V., Sriramkumar, M., Jothivenkatachalam, K., Saravanan, P., Rajamohan, N., Vasseghian, Y., Rajasimman, M., 2022. Impact of Erbium (Er) and Yttrium (Y) doping on BiVO4 crystal structure towards the enhancement of photoelectrochemical water splitting and photocatalytic performance. *Chemosphere* 299, 134343. https://doi.org/10.1016/j.chemosphere.2022.134343

Noor, S., Ashar, A., Taj, M.B., Bhutta, Z.A., 2022. Advanced oxidation processes for remediation of persistent organic pollutants. In: *Advanced Oxidation Processes for Wastewater Treatment* (pp. 203–212). CRC Press. https://doi.org/10.1201/9781003165958

Novotna, K., Cermakova, L., Pivokonska, L., Cajthaml, T., Pivokonsky, M., 2019. Microplastics in drinking water treatment–current knowledge and research needs. *Sci. Total Environ.* 667, 730–740. https://doi.org/10.1016/j.scitotenv.2019.02.431

Pandian, A.M.K., Gopalakrishnan, B., Rajasimman, M., Rajamohan, N., Karthikeyan, C., 2021. Green synthesis of bio-functionalized nano-particles for the application of copper removal–characterization and modeling studies. *Environ. Res.* 197, 111140. https://doi.org/10.1016/j.envres.2021.111140

Parsons, S. (Ed.). 2004. *Advanced Oxidation Processes for Water and Wastewater Treatment.* IWA Publishing.

Patil, P.N., Gogate, P.R., 2015. Degradation of dichlorvos using hybrid advanced oxidation processes based on ultrasound. *J. Water Proc. Eng.* 8, 58–65. https://doi.org/10.1016/j.jwpe.2014.10.012

Saharan, V.K., Pinjari, D.V., Gogate, P.R., Pandit, A.B., 2014. Advanced oxidation technologies for wastewater treatment: An overview. *Indust. Wastewater Treat. Recycl. Reuse*, 141–191. https://doi.org/10.1016/j.ultsonch.2020.105306

Sharma, N.K., Philip, L., 2016. Combined biological and photocatalytic treatment of real coke oven wastewater. *Chem. Eng. J.* 295, 20–28. https://doi.org/10.1016/j.cej.2016.03.031

Sun, S., Jiang, T., Lin, Y., Song, J., Zheng, Y., An, D., 2020. Characteristics of organic pollutants in source water and purification evaluations in drinking water treatment plants. *Sci. Total Environ.* 733, 139277. https://doi.org/10.1016/j.scitotenv.2020.139277

Wan, L., Wang, H., 2021. Control of urban river water pollution is studied based on SMS. *Environ. Technol. Innovat.* 22, 101468. https://doi.org/10.1016/j.eti.2021.101468

Xia, J., Zhang, Y., Xiong, L., He, S., Wang, L., Yu, Z., 2017. Opportunities and challenges of the Sponge City construction related to urban water issues in China. *Sci. China Earth Sci.* 60, 652–658. https://doi.org/10.1007/s11430-016-0111-8

Yahya, M.S., Oturan, N., Kacemi, K. El, Karbane, M., Aravindakumar, El C.T., Oturan, M. A., 2014. Oxidative degradation study on antimicrobial agent ciprofloxacin by electroFenton process: Kinetics and oxidation products. *Chemosphere* 117, 447–454. https://doi.org/10.1016/j.chemosphere.2014.08.016

Yang, Z., Zou, L., Xia, J., Qiao, Y., Bai, F., Wang, Q., Cai, D., 2022. Spatiotemporal variation characteristics and source identification of water pollution: Insights from urban water system. *Ecol. Indicat.* 139, 108892. https://doi.org/10.1016/j.ecolind.2022.108892

Zhang, Y., Wang, H., Li, Y., Wang, B., Huang, J., Deng, S., Yu, G., Wang, Y., 2020. Removal of micropollutants by an electrochemically driven UV/chlorine process for decentralized water treatment. *Water Res.* 183, 116115. https://doi.org/10.1016/j.watres.2020.116115

8 Microstructure Characterization and Characterization Techniques of Adsorbent Materials

Diya Kuttappa, Sayonjit Roy, Rounak Ghosh, and Ishita Auddy

8.1 INTRODUCTION

Adsorption is defined as the deposition of an adsorbate (solid/gas) on the surface of an adsorbent (solid) used to eliminate contaminants from liquid phases. The adsorption phenomenon, since its discovery over a century ago, has been used in numerous applications, such as air purification, nutrient recovery, chemical detection, humidity removal, fuel purification, water purification, wastewater treatment, and solar energy conversion, for a few examples. The mass transfer process of adsorption is driven by the differential in concentration of the adsorbate in the fluid (liquid or gas) and solid phases (adsorbent surface) [1]. Adsorbents are solid materials that are highly porous or a bed or coating of tiny particles. Because of their large number of micropores, they frequently have very large specific surface areas. The high specific surface areas of adsorbents offer a multitude of possibilities for the adsorption process to absorb harmful or undesirable compounds [2]. The type of adsorption influences the bonding energy between molecules and surface, chemical adsorption requires a higher energy than physical adsorption.

> Activated carbons (AC), carbon nanotubes, carbon molecular sieves, activated alumina, pillared clays, silica gels, synthetic polymers, zeolites, aluminosilicate zeolites, zirconia, and metal–organic frameworks (MOFs) are the most prevalent forms of porous adsorbent materials [8].

Many separations that are neither achievable nor viable using traditional methods, such as distillation, absorption, and even membrane-based systems, can be accomplished by adsorption. Adsorption has seen a significant expansion in applications recently due to dramatically increasing environmental or quality constraints.

DOI: 10.1201/9781003767251-8

Likewise, many of those expectations can now be satisfied because of developments in absorbent technology. Adsorbents that have recently been produced are now "off-the-shelf" and may typically function satisfactorily. However, novel adsorbents with significantly enhanced qualities that result in increased performance are always being created [7].

> One of the most successful physicochemical techniques for treating textile effluent is adsorption. The development of alternatives to activated carbon, the expensive and traditional adsorbent advised for the removal of dyes from colored effluents, is gaining traction [4]. Owing to the intriguing properties of the adsorbent materials, which include a high specific surface area, a well-developed porous structure, a high adsorption capacity, and rich surface chemistry that can be adjusted by altering the production process all the way up to the application's end.

> The agricultural wastes made of cellulosic components are today considered cheap, plentiful, and easily accessible natural resources for the biochar, paper, and chemical industries [3]. Traditional adsorbents have limitations such as rigidity and anionic nature of aluminosilicate frameworks, high production costs, lack of homochirality, and challenges with functionalization. The most often used adsorbents are ACs due to their affordable cost and widespread availability in the market [8].

Mesopores usually have a radius of 1 and 25 nm, which have a radius of less than 1 nm, and macropores have a radius of more than 25 nm, which are the three types of pores found in adsorbent material throughout the solids. Adsorbent materials like carbons and silica gels are made of intricate, amorphous networks of micropores, mesopores, and macropores that are interconnected.

8.1.1 Types of Adsorbent Materials

These substances are grouped into three types such as organic, inorganic, and hybrid. They differ in terms of membranes and particle sizes, ranging from nano to micro. Adsorbents can come from a variety of sources, including synthetic and natural materials that have been altered chemically, physically, or both, as well as structures and material combinations. Activated carbon, zeolites, siliceous materials, carbon nanomaterials (graphene, carbon nanotubes), biomass/biochar, MOFs, clay materials (kaolinite, bentonite), polymers, etc., are some of the materials that are broadly applied [1, 4]. Commercially accessible adsorbents come in a variety of forms, and choosing the right one from the wide range of materials available can frequently be challenging for the user. For a precise measurement of volatile organic compounds (VOCs), the adsorbent material must fulfil the following requirements:

- Full enrichment of the target analytes.
  The specific surface area and the porous structure provide a ballpark estimate of a material's adsorption strength. The particular breakthrough volume of model compounds provides a closer characterization of the adsorption strength.

- Analytes desorb completely and rapidly.
- An even and neutral surface to prevent the creation of artefacts, permanent adsorption, and catalytic effects during sampling, storing the filled adsorbent tubes, and desorption.
- Low ability to adsorb other inorganic air elements as ozone, carbon dioxide, sulfur dioxide, nitrogen oxide, and so forth.
- High resistance to reactive species, such as ozone.
- Numerous applications.
- Excellent mechanical and thermal stability [9].

8.1.2 Adsorbent Granules

Absorbent grains are loose, versatile absorbents that are used for packing laboratories, cleaning up spills, and stabilizing free liquids. Although the granule size is reduced, the contact area between the granules is improved, thereby raising the mass transfer resistance and, ultimately, the rate of heat transmission through the bed. Improving the heat and mass transfer capacity of a granular adsorbent bed requires an understanding of the mechanisms underlying the mass and heat transfer in the bed.

In certain cases, liquids can be absorbed from the voids and fissures of a range of surfaces using granular absorbent [10].

Wet granulation refers to a technique in which there is an enlargement in the size of granules by the action of rolling, tumbling, and shear forces, which force the raw material to form granules by the agglomeration mechanism. The choice of binder for adsorbent granules remains a significant step since the binder must be stable and maintain the granules' structure in aqueous solution throughout the adsorption process. Compost granules have a comparable adsorption capability to other adsorbent materials, indicating that they have a great deal of potential for application as an absorbent material [11].

8.1.3 Adsorbent Socks/Pillows

Large volumes of both hazardous and non-hazardous liquids, such as oil, water, acid, bases, and many other liquids, can be absorbed by absorbent chemical pillows. Unknown spills are most effectively adsorbed using these pillows. These are usually made of polypropylene (PP). These are made to take up massive volumes of oil and oil products that would be too much for a pad to handle. These cushions are excellent for a variety of tasks, particularly collecting spills and drips beneath machinery. It works well with all kinds of oils, solvents, water, coolants, and other elements [12, 13, 14].

8.1.3.1 Importance of Characterization of Adsorbent Materials

Adsorption capacity (or 'loading') is the most important characteristic of an adsorbent. It is stated as- 'the amount of adsorbate taken up by the adsorbent, per unit mass (or volume) of the adsorbent'. A thorough material characterization is essential when a new adsorbent is offered in order to comprehend its performance in the intended process, support its application, and know its properties. Depending on the kind of material to be researched and the equipment available, many procedures can be used to collect a variety of physical and chemical information. UV-vis, FTIR, Raman, XPS, EDX, zeta potential, XRD, SEM/FESEM, TEM, AFM, surface area and porosimetry, VSM, DLS, and thermoanalytical analysis are a few of the methods that are available [15, 16, 17]. These methods make it possible to examine the structure, topography, morphology, and magnetic properties of materials in addition to their other physical characteristics. They also make it possible to identify, locate, and quantify chemical constituents such as elements, functional groups, and molecules [1].

In order to give details on the framework, anatomy, topography, strength, composition, and dimension of the adsorbent material, as well as other crucial qualities for the adsorption processes, research including its characterization is crucial [18]. This information makes it easier to plan new adsorbents, evaluate, select adsorbents, and put adsorption devices and processes into practice in the real world [1].

8.1.3.2 Chemical Characterization

Adsorbent materials can be chemically characterized using a range of methods that offer vital details about their chemical structure, including identification, location, and quantification. Beyond revealing details on the make-up of adsorbents, this kind of characterization can also shed light on the essential categories that can be observed on the material's surface and are connected to the adsorption processes, such as hydrogen bonding [1, 5].

8.1.3.3 Physical Characterization

When evaluating an adsorbent material, some of the most important physical attributes to consider include topography, morphology, size, magnetic properties, and structure. X-ray diffraction (XRD) analysis, surface area and porosimetry analysis, dynamic light, atomic force microscopy (AFM), scanning electron microscopy (SEM), transmission electron microscopy (TEM), field-emission scanning electron microscopy (FESEM), and vibrating sample magnetometer (VSM) techniques are used for material characerization [5, 6].

8.1.3.4 Thermal Characterization

A material's or reaction product's physical characteristics can be assessed as a function of temperature using a variety of thermal analysis techniques. The materials under evaluation are placed in a designated atmosphere and are subjected to a controlled temperature program. These analytical techniques, which include differential thermal analysis (DTA), thermogravimetry (TG), and differential scanning, are based on the evaluation of a physical attribute as a function of temperature [5].

8.2 DIFFERENT TECHNIQUES FOR THE CHARACTERIZATION OF ADSORBENT MATERIAL

8.2.1 Microstructural Analysis

Microstructural analysis plays a crucial role in characterizing adsorbent materials. The branch of materials science known as "microstructural analysis" examines a material's microscopic structure. The arrangement of atoms, crystals, grains, and phases within a material can be understood by this examination. The characteristics and functionality of a material are greatly influenced by its microstructure [19, 20, 21].

8.2.2 Microscopy Techniques

Scanning Electron Microscopy (SEM): High-resolution pictures of the surface morphology of adsorbent materials may be obtained using SEM. It provides information on the characteristics of the surface, shape, and size of particles [33].

Transmission Electron Microscopy (TEM): TEM allows for even higher resolution imaging, providing details on the internal structure and crystallinity of adsorbents at the nanoscale level. A good reference for TEM techniques is "Transmission Electron Microscopy" [22, 23].

Atomic Force Microscopy (AFM): AFM offers topographic imaging and measures surface forces at the atomic scale, useful for understanding surface roughness and interaction forces [34].

Scanning Tunneling Microscopy (STM): STM visualizes surfaces at the atomic scale, aiding in understanding surface defects and electronic properties. "Scanning Tunneling Microscopy and its Applications" by C. Julian Chen is a comprehensive reference for STM [34].

8.2.3 X-ray Techniques

In the area of microstructural investigation, X-ray techniques are effective tools that yield important details regarding the composition and crystallography of materials. The following are some essential, often-used X-ray methods:

- **X-ray Diffraction (XRD)**: One method for examining a material's crystal structure is XRD. A crystalline sample experiences diffraction when X-rays interact with it, creating a distinctive pattern. It is possible to identify the crystalline phases present in the material as well as determine the crystal structure and orientation by analyzing this diffraction pattern [33].
- **X-ray Powder Diffraction (PXRD)**: PXRD is an XRD variant made especially for materials that are powdered. It is frequently employed in the quantitative examination of crystalline components in a sample, the identification of crystalline phases, and the assessment of crystal structure [35].

- **Fluorescence X-ray (XRF)**: A non-destructive method for analyzing a material's elements is XRF. Depending on its elemental make-up, a sample that is subjected to high-energy X-rays will produce distinct X-ray fluorescence. The elemental composition and concentration of the sample can be ascertained by measuring the X-rays that are released [35].
- **Photoelectron Spectroscopy, also known as XPS or ESCA**: X-ray photoelectron spectroscopy (XPS) is a surface-sensitive technology that offers details on the chemical state and elemental composition of a material's topmost layers. It entails exposing a sample to X-ray radiation and calculating the kinetic energy of the photoelectrons released [36].

8.3 PORE SIZE DISTRIBUTION ANALYSIS

Analyzing the distribution of pore sizes inside a material is essential for describing porous materials because it sheds light on the variety of pore sizes that are present and how they are distributed. Numerous disciplines, such as materials science, geoscience, and environmental science, depend on an understanding of pore size distribution. Many techniques are used to examine the distribution of pore size [37].

8.3.1 Mercury Intrusion Porosimeter

An advanced tool for analyzing the pore structure of solid materials is a mercury intrusion porosimeter (MIP). Mercury porosimetry, another name for this method, is very useful for figuring out the overall pore volume and pore size distribution in wide range of components, namely metals, plastics, glasswares, and geological samples. The basic idea behind MIP is that a material's pores might allow mercury to seep into it [38]. The reason mercury was selected is that it is a non-wetting liquid, which means it does not contaminate the surfaces of solids. Because of this, it can enter pores unaffected by surface tension [38].

The sample is put inside the porosimeter chamber; it is frequently in the shape of a solid compact or powder. A high-pressure system built into the device enables the carefully measured addition of mercury to the sample. Under increasing pressure, mercury is injected into the sample. Continuous measurements and records are made of the amount of mercury that has seeped into the substance [39]. The size of the holes determines the pressure at which mercury seeps into the substance. The device can produce a pore size distribution curve, which shows the range of pore sizes present in the material, by evaluating the pressure-volume data [39].

8.3.2 Gas Adsorption

An essential method for examining a material's surface area and porosity is gas adsorption, which is especially useful for porous materials like activated carbon, zeolites, and catalysts. The procedure is calculating how much gas is adsorbed onto a solid material's surface in relation to concentration or pressure [40]. The surface area, pore size distribution, and other important characteristics of the material are then ascertained using this information. A solid material's surface experiences gas molecules being adsorbed onto it. Usually, adsorption is a physical process in which van der Waals forces or other attractive forces cause gas molecules to stick to the surface [41].

An adsorption isotherm describes the relationship between the amount of gas adsorbed and the gas's pressure or concentration.

Carbon dioxide (CO_2), argon (Ar), and nitrogen (N_2) are often utilized gases in adsorption investigations [43]. One popular approach for calculating specific surface area from gas adsorption data is the BET method [42].

Assuming that a homogeneous layer of gas molecules occurs on the surface, it is predicated on the study of monolayer adsorption [42]. On the surface, gas adsorption might take place in several layers. One can learn about the pore size distribution of the material by examining the adsorption at various pressures [43].

8.4 SURFACE CHEMISTRY ANALYSIS

8.4.1 Fourier-Transform Infrared Spectroscopy (FTIR)

A significant analytical method for determining and characterizing a material's chemical composition based on how it interacts with infrared (IR) light is FTIR. The functional groups, chemical bonds, and molecular structure of a sample are all revealed by FTIR spectroscopy [44].

The fundamental principle of FTIR is the belief that specific infrared radiation waves are taken up by particles, causing the bonds between the atoms to vibrate differently. A wide range of infrared light is applied to the sample, and the absorption of various wavelengths is measured. The far-infrared (FIR) and near-infrared (NIR) are the two main spectral ranges that are covered by FTIR [44].

Information regarding various chemical vibration types can be found in different parts of the spectrum. Many different fields, such as chemistry, biology, pharmaceuticals, polymers, and materials research, use FTIR extensively. It is used for chemical reaction monitoring, identification of unknown substances, and qualitative and quantitative analysis [43].

8.4.2 X-ray Spectroscopy (XPS)

X-ray Electron Spectroscopy for Chemical Analysis (ESCA), also referred to as Photoelectron Spectroscopy (XPS), is a sophisticated surface analysis method used to describe the electronic structure, chemical state, and elemental make-up of materials. The outermost layers of a sample, which are usually a few nanometers deep, can be usefully revealed by XPS [33]. The photoelectric effect, which is the basis of XPS, uses X-ray photons to eject electrons from the innermost cavities of the sample's molecules. It is possible to detect the kinetic energy of expelled electrons and, in turn, their binding energy [34]. Electron energy analyzers are used to measure the electrons released during photoionization.

The photoelectrons' kinetic energy, which is correlated with their binding energy, is measured by the analyzer. Surface chemistry, materials science, and many other industries use XPS extensively [46]. Thin film, coatings, polymers, semiconductors, catalysts, and biological sample surfaces are among the materials it is used to analyze. Quantitative data regarding a sample's elemental composition can be obtained via XPS [46].

Quantitative analysis makes use of variables like the sensitivity factors for various elements and the strength of XPS peaks. X-ray electron photoelectron spectroscopy is a non-destructive method that is essential to comprehending material characteristics and surface chemistry [47]. It is an effective instrument for study and development in a variety of scientific and commercial fields since it can give information on both elemental and chemical states [46].

8.5 MECHANICAL PROPERTIES CHARACTERIZATION

8.5.1 Compression Testing

A mechanical assessment technique called compression testing is used to assess how materials will behave under compressive stresses. It is an essential test in materials science and engineering that gives important details regarding the compressive strength, deformation properties, and axial load tolerance of a material. A compressive force is applied to a specimen along its longitudinal axis during a compression test [27, 28, 29].

The material's response to the applied force, including its deformation and eventual failure under compression, is measured during the test. A testing apparatus that can impart controlled axial loads to the specimen could be hydraulic or electromechanical and is a standard component of a compression testing system. Usually shaped like a cube or cylinder, the specimen's dimensions are precisely determined before testing [30, 31, 32].

8.6 TENSILE TESTING

A common mechanical testing technique used to assess the mechanical characteristics of components under axial tensile strain is tensile testing, also referred to as tension testing. This type of testing yields crucial information about a material's behavior under stress, including its strength, elasticity, ductility, and deformation characteristics. [48].

In tensile testing, an axial load is applied to a specimen, and the applied force is progressively increased until the material deforms or fails [49]. By calculating the connection between the applied force and the resulting deformation, the test analyzes the mechanical characteristics of the substance. Specimens are prepared in accordance with accepted practices, with particular measurements determined by the material and testing specifications [48].

Precise testing outcomes and consistency are guaranteed by meticulous machining or preparation. Tensile testing yields information regarding the ductility of the material, which is defined as its capacity to undergo plastic deformation prior to breaking. Commonly used metrics for ductility include percentage elongation and area decrease.

Many industries, including manufacturing, aerospace, automotive, and construction, use tensile testing extensively [49]. It supports material selection, quality assurance, and structural and product design. Ensile tests can be carried out in a variety of settings, for example, at room temperature or at a higher temperature [50].

Different loading rates may also be used in tests to examine how the material behaves in dynamic situations. A basic and popular technique for comprehending the mechanical characteristics of materials is tensile testing [49]. When designing and engineering materials for applications where their strength and deformation characteristics are crucial considerations, the outcomes of tensile testing are crucial [50].

8.7 WATER ABSORPTION ANALYSIS

8.7.1 Gravimetric Method

Absorbent materials play a crucial role in various industries, ranging from healthcare to environmental protection. Understanding their water absorption properties is essential for optimizing their performance. Among the various characterization techniques available, the gravimetric method stands out as a reliable and widely used approach. This review explores the significance of the gravimetric method in water absorption analysis, its principles, applications, and recent advancements [45]. Through an examination of relevant literature, our goal is to present a thorough grasp of the ways in which this technique aids in the characterization of absorbent materials. [26]. Absorbent materials are substances capable of taking in and retaining liquid, making them vital in applications such as medical products, hygiene items, environmental protection, and beyond. Efficient characterization of these materials is essential to ensure their optimal performance. Among the various techniques available, the gravimetric method has gained prominence for its accuracy and versatility in water absorption analysis.

8.7.2 Principles of Gravimetric Method

The gravimetric method relies on the measurement of weight changes in a material as it absorbs water. This technique involves exposing the absorbent material to a controlled environment with a known humidity level or immersing it in a liquid. The weight gain of the material is then monitored over time, providing valuable information about its water absorption properties. The fundamental principle underlying the gravimetric approach is predicated on the idea that the material's mass increases as it absorbs water molecules. By precisely measuring the change in mass, researchers can calculate the water absorption capacity of the material [24]. This method offers high sensitivity and accuracy, making it a preferred choice in absorbent material characterization.

8.7.3 Applications of Gravimetric Method in Water Absorption Analysis

- **Hygiene Products**: Gravimetric analysis is extensively employed in the characterization of materials used in hygiene products such as diapers, sanitary napkins, and adult incontinence products. Understanding the water absorption capabilities of these materials is crucial for designing products that provide optimal comfort and performance.

- **Medical Devices**: Absorbent materials are utilized in various medical devices, including wound dressings and surgical products. Gravimetric analysis helps assess the water absorption kinetics of these materials, ensuring that they meet the specific requirements for medical applications.
- **Environmental Protection**: Absorbent materials are essential in environmental protection applications, such as oil spill clean-up materials. Gravimetric analysis aids in evaluating the effectiveness of these materials in absorbing and retaining liquids, contributing to environmental remediation efforts.
- **Agriculture**: Agriculture benefits from absorbent materials in products like soil conditioners and water-retaining polymers. Gravimetric analysis allows researchers to understand how these materials interact with water, influencing soil moisture levels and plant growth.

Recent research has focused on enhancing the capabilities of gravimetric analysis for water absorption characterization. One notable advancement involves the integration of automated systems and advanced sensors, allowing for real-time monitoring of water absorption kinetics. This enables researchers to obtain more detailed and precise data, improving the understanding of absorbent material behavior [25].

8.7.4 Swelling Ratio Measurement

Absorbent materials play a pivotal role in various industries, from medical applications to environmental remediation. Understanding the water absorption behavior of these materials is essential for optimizing their performance in specific applications. Swelling ratio measurement is a critical aspect of water absorption analysis, providing valuable insights into the structural and morphological changes of absorbent materials. This review paper aims to comprehensively explore the characterization techniques employed for assessing water absorption in absorbent materials, with a focus on swelling ratio measurement. Absorbent materials, ranging from polymers to natural substances, find widespread applications in fields such as healthcare, agriculture, and environmental engineering. The ability of these materials to absorb and retain water is a key parameter that directly influences their functionality. Water absorption analysis, particularly through swelling ratio measurement, is an indispensable tool for characterizing absorbent materials.

8.7.5 Swelling Ratio Measurement

Swelling ratio, defined as the volume increase of a material upon water absorption, is a crucial factor to consider while analyzing absorbent elements. This measurement is crucial for understanding the material's response to water and provides insights into its structural changes. Various techniques are employed to measure the swelling ratio, including gravimetric methods, optical techniques, and imaging analyses.

- **Gravimetric Methods**: Gravimetric methods involve measuring the weight change of a material as it absorbs water. The initial and final weights are recorded, and the swelling ratio is calculated based on the volume change. This traditional approach is simple and widely used, providing accurate results for a broad range of absorbent materials. However, it may not capture the dynamic changes occurring during the absorption process.
- **Optical Techniques**: Optical techniques, such as spectrophotometry and turbidimetry, offer non-destructive ways to monitor swelling in real time. These methods utilize the interaction of light with the material during water absorption. Spectrophotometric measurements involve analyzing changes in the material's optical density, while turbidimetry assesses the scattering of light. Optical techniques are advantageous for their high sensitivity and ability to capture dynamic changes, making them suitable for time-dependent swelling studies.
- **Imaging Analyses**: Advanced imaging techniques, including SEM and AFM, provide detailed insights into the morphological changes of absorbent materials during water absorption. These methods allow for visualizing surface alterations, porosity, and structural deformations at a microscopic level. Imaging analyses complement other swelling ratio measurement techniques, offering a thorough comprehension of the behavior of the item.

8.7.6 Characterization Techniques

In addition to swelling ratio measurement, a range of characterization techniques is employed to assess absorbent materials comprehensively. These techniques include FTIR, XRD, and NMR.

- **Fourier-Transform Infrared Spectroscopy (FTIR)**: FTIR is utilized to analyze the chemical composition of absorbent materials before and after water absorption. The technique provides information about functional groups, allowing researchers to identify changes in molecular structure. Understanding these changes is essential for optimizing absorbent materials for specific applications.
- **X-ray Diffraction (XRD)**: XRD is a technique utilized to examine the crystalline arrangement of absorbent materials. Water absorption can induce structural changes, leading to alterations in crystallinity. XRD helps in identifying these changes, providing valuable information about the material's stability and performance under different moisture conditions.
- **Nuclear Magnetic Resonance (NMR)**: NMR is a powerful technique for studying molecular dynamics in absorbent materials. By examining the interaction between water molecules and the material's structure, NMR can reveal insights into the diffusion and distribution of water within the material. This information is crucial for optimizing absorbent materials for controlled water release applications.

8.8 MICROSTRUCTURE–PROPERTY CORRELATION

8.8.1 Linking Microstructure to Performance

Extremely absorbent substances are crucial in a broad spectrum of applications, including environmental clean-up and medical equipment. Gaining a comprehensive understanding of the microstructure of these materials is essential for maximizing their performance. This text presents an outline of sophisticated methods used to analyze the microstructure of absorbent materials and explores the ways in which these methods improve their performance. The mentioned study and review paper underscore the significance of microstructure analysis in the advancement and refinement of absorbent materials. Absorbent materials, or sorbents, are compounds capable of absorbing and retaining liquids or gases. They are widely used in several domains, including environmental clean-up, healthcare, and industrial operations. The performance of absorbent materials is closely tied to their microstructure, which includes the organization, distribution, and characteristics of the component constituents at the microscopic scale. Comprehending the microstructure of absorbent materials is crucial for customizing their characteristics to suit certain uses. This research examines the correlation between absorbent fabrics' microstructure and characteristics and emphasizes the need of improved characterization techniques to understand the underlying processes.

8.8.2 Microstructure–Property Correlation in Absorbent Materials

The qualities of absorbent materials, such as porosity, surface area, and mechanical strength, are directly affected by their microstructure. Porosity is a critical factor in influencing the ability of a material to absorb substances. A clearly defined microstructure offers valuable information about the possible routes for fluid infiltration and retention inside the material's matrix. The microstructure of absorbent materials is closely associated with their mechanical strength, which is a crucial feature. The spatial organization of fibers, particles, or polymers at the microscopic scale impacts the overall structural robustness and longevity of the material. Comprehending these connections is crucial for developing absorbent materials that can endure all the challenges presented by specific applications.

8.8.3 Characterization Techniques for Microstructure Analysis

Different sophisticated techniques are utilized to examine the microscopic composition of absorbent materials. These approaches provide a comprehensive understanding of the structure, form, and arrangement of constituents inside the material. Notable methods for characterizing include the following:

- **Scanning Electron Microscopy (SEM)**: SEM offers detailed pictures with high resolution that depict the surface morphology of absorbent materials. It is highly advantageous for seeing the microstructure, discerning the distribution of pore sizes, and noticing any alterations to the surface.

- **Transmission Electron Microscopy (TEM)**: TEM enables the investigation of the interior microstructure at a nanoscale resolution. This approach is crucial for examining the organization of nanoparticles or nanofibers within absorbent materials.
- **X-ray Diffraction (XRD)**: XRD is used to examine the crystal structure of materials that can absorb substances. It assists in the identification of various stages, crystallinity, and crystallographic orientation, providing valuable information regarding physical and thermodynamic characteristics of the material.
- **Nuclear Magnetic Resonance (NMR)**: NMR is a powerful technique for investigating the molecular dynamics and interactions occurring inside absorbent materials. It offers insights into the movement of water molecules, facilitating comprehension of the material's ability to absorb.
- **BET Surface Area Analysis**: This technique quantifies the precise surface area of materials by measuring the adsorption of gas. Understanding the material's porosity and surface properties is essential since it directly affects its absorption and adsorption capacities.

8.8.4 Case Studies and Applications

This part highlights case studies that illustrate the real-world effects of microstructure–property correlation. These studies emphasize the effective utilization of sophisticated characterization techniques to optimize absorbent materials for specific applications.

- **Environmental Remediation**: Absorbent materials are crucial in the process of environmental cleaning, namely in the extraction of contaminants from water or air. Researchers can boost the effectiveness of absorbing and eliminating certain impurities by designing absorbents that correlate the microstructure with the required qualities.
- **Biomedical Applications**: Absorbent materials are employed in medical contexts for wound dressings, medication delivery systems, and diagnostic instruments. By customizing the microstructure of these materials, their performance may be optimized, leading to enhanced tissue regeneration, precise medication release control, and accurate sensing abilities.
- **Industrial Processes**: Absorbent materials are utilized in many industrial processes, such as gas separation and purification. Comprehending the microstructure enables the creation of materials that possess enhanced selectivity and capacity for certain gases, hence augmenting the efficiency of industrial applications.

8.8.5 Designing Tailored Adsorbent Materials

Within the domain of materials science, there has been considerable focus on the research and development of adsorbent materials. These materials play a crucial role in a range of industrial applications, specifically in fields such as environmental clean-up, gas separation, and water purification. This research examines the complex connection between the microstructure and characteristics of adsorbent materials, highlighting the significance of customized designs in improving their performance. Furthermore, an extensive examination of sophisticated characterization techniques provides insights into the strategies used to comprehend and enhance these materials. Gaining a clearer comprehension of the basic ideas underlying the construction of adsorbent materials is the aim of this study. Specifically, it intends to investigate the relationship between the microstructure of these materials and their characteristics. The findings of this study will serve as a guide for future advancements in this quickly developing sector.

8.8.5.1 Tailored Microstructures for Enhanced Adsorption Properties: Pore Dimensions Dispersion and Surface Area

The microstructure, namely the distribution of pore sizes and the extent of surface area, is a crucial element that determines the adsorption capacity of materials. Research has demonstrated that customizing these characteristics may greatly influence the ability of materials to adsorb substances. For example, mesoporous materials that have clearly defined pore architectures are shown to have excellent abilities to adsorb certain contaminants. Previous research has investigated the effective design of mesoporous silica materials for the efficient extraction of heavy metal ions from water-based solutions.

8.8.5.1.1 Functional Groups and Chemical Composition

Applying specific functionalities to the surface of adsorbent materials improves their ability to selectively capture target contaminants. Controlling the chemical composition of the material is a crucial factor in designing its microstructure. For example, the addition of amine groups has been found to be useful in improving the ability of a substance to attract and hold CO2 in gas separation processes.

8.8.5.2 Microstructure–Property Correlation

Recognizing the connections between microstructure and characteristics is essential for the logical development and enhancement of adsorbent materials. This section examines several facets of this association, encompassing the impact of pore size, surface chemistry, and structural flaws on the effectiveness of adsorption.

8.8.5.2.1 Impact of Pore Size on Adsorption Capacity

The dimensions of pores in adsorbent materials are crucial in influencing their adsorption capability. The adsorption behavior of gases is substantially influenced by the relative kinetic diameter of the molecules in relation to the pore size. The study conducted by Li et al. (2019) demonstrates that manipulating the diameters of pores at various length scales can result in improved ability to adsorb a variety of gas molecules in hierarchical porous materials.

Surface chemistry refers to the study of chemical reactions and processes that occur at the interface between two phases, such as a solid and a gas or a solid and a liquid. Selectivity, on the other hand, refers to the ability of a catalyst or a chemical reaction to prefer. The interaction between adsorbates and a material's surface is determined by the chemical composition of the surface. Modifying the surface chemistry enables the targeted adsorption, which is especially vital in tasks like water purification and gas separation. The development of MOFs that possess distinct functional groups in order to attain exceptional selectivity in gas adsorption.

Manipulating the structure of adsorbent materials can have an impact on the rate at which adsorption occurs. Gaining insights into the relationship between flaws and adsorption dynamics is crucial for maximizing the activity of materials in terms of fast and effective adsorption. The study conducted examines how defect engineering affects the rate at which organic pollutants are adsorbed. This research offers useful insights into the development of materials with a high concentration of defects, which can lead to improved performance.

8.8.5.3 Advanced Methods for Analyzing and Understanding the Characteristics of Something

Advanced characterization techniques are essential for understanding the complex relationship between microstructure and property. This section presents a comprehensive review of advanced techniques used to analyze adsorbent materials. SEM and TEM are two types of microscopy techniques.

Scanning electron microscopy (SEM) and transmission electron microscopy (TEM) provide researchers with the ability to observe the detailed structure and composition of adsorbent materials at the nanoscale. These approaches are quite important for evaluating the consistency of pore distribution and comprehending the overall structure of the material.

X-ray diffraction (XRD) is an effective method used to determine the crystalline structure of adsorbent materials. Through the examination of diffraction patterns, scientists are able to determine the types of crystal structures present, evaluate the size of the individual crystals, and obtain information on the overall structural stability of the material [33].

NMR spectroscopy: Nuclear magnetic resonance (NMR) spectroscopy offers precise insights into the specific surroundings of atoms inside a substance. This method is especially valuable for investigating the surface chemistry of adsorbent materials, which involves determining the presence of functional groups and examining how they interact with adsorbates. Adsorption isotherms and thermogravimetric analysis (TGA) plays a vital part in deciding the adsorption capacity and thermal stability of compounds. These approaches yield significant data about the behavior of adsorbent materials in terms of adsorption–desorption and their stability, which is influenced by temperature.

8.9 FUTURE ASPECTS

As we progress into the future, the demand for efficient and sustainable materials continues to grow, particularly in fields such as environmental science, energy, and healthcare. One area that holds immense promise is the microstructure characterization and characterization techniques of absorbent materials. These materials play a pivotal role in various applications, from environmental remediation to medical technologies. Understanding their microstructure and employing advanced characterization techniques can unlock new possibilities and enhance the performance of these materials. In this comprehensive review, we will delve into the current state of research, identify gaps, and propose future directions for exploration.

8.9.1 Current State of Research

Microstructure characterization is crucial for unravelling the intricate details of absorbent materials, as it directly influences their properties and performance. Researchers have made significant strides in employing various techniques to investigate the microstructure of absorbent materials. Furthermore, the development of advanced imaging techniques, such as three-dimensional (3D) imaging and tomography, has allowed researchers to explore the internal structures of absorbent materials in unprecedented detail. These techniques provide valuable insights into pore size distribution, surface morphology, and interconnectivity, which are crucial factors determining the efficiency of absorbent materials.

8.9.2 Characterization Techniques

- Advanced Imaging Techniques:
  - 3D Imaging and Tomography: The use of 3D imaging and tomography allows researchers to visualize the internal structure of absorbent materials in three dimensions. This technique can provide detailed information about the spatial distribution of pores, enhancing our understanding of material behavior.
- Spectroscopic Techniques:
  - Nuclear Magnetic Resonance (NMR): NMR spectroscopy is a powerful tool for studying the molecular dynamics and interactions within absorbent materials. By analyzing the NMR spectra, researchers can gain information about the molecular mobility and chemical make-up of the substance.
- Microscopy Techniques:
  - Scanning Electron Microscopy (SEM): SEM is widely used for surface imaging and provides high-resolution images of the material's morphology. Combining SEM with energy-dispersive X-ray spectroscopy (EDS) allows for elemental analysis, offering a comprehensive understanding of the material's composition.
- X-ray Techniques:
  - X-ray Diffraction (XRD): XRD is instrumental in determining the crystal structure and phase composition of absorbent materials. By analyzing XRD patterns, researchers can identify specific crystalline phases, aiding in the optimization of material synthesis.

8.9.3 Future Avenues of Research

- Multiscale Characterization: Future research should focus on integrating multiscale characterization techniques to understand the hierarchical structure of absorbent materials. Combining macroscopic imaging with nanoscale investigations will provide a holistic view of the material, allowing for tailored design and improved performance.
- In situ Characterization: The development of in situ characterization techniques will enable researchers to observe dynamic changes in absorbent materials under real operating conditions. This is particularly crucial for applications such as water purification and drug delivery, where the material's performance is influenced by external factors.
- Machine Learning Applications: Incorporating machine learning algorithms for data analysis can enhance the efficiency of microstructure characterization. By training algorithms on vast datasets obtained from various characterization techniques, researchers can quickly identify patterns and correlations that might be challenging for manual analysis.
- Functionalization and Tailoring: Research efforts should explore the functionalization of absorbent materials to enhance their specific properties. This includes surface modifications, doping with nanoparticles, and engineering materials at the molecular level to achieve targeted absorbent capabilities for specific applications.
- Environmental Applications: The future of absorbent materials lies in their sustainable applications. Research should aim to develop eco-friendly absorbent materials for environmental remediation, such as removing pollutants from water and air. Understanding the microstructure of these materials is crucial for designing efficient and sustainable solutions.

8.10 CONCLUSION

The evaluation of microstructure is crucial for comprehending the functionality and properties of absorbent materials, exerting a significant impact on several fields such as environmental remediation and medical devices. This extensive investigation has examined numerous aspects of microstructure analysis and the corresponding approaches used for absorbent materials. The knowledge acquired from these studies greatly enhances the development, refinement, and progress of absorbent materials for various applications. An important discovery from these research is the significant influence of microstructure on the effectiveness of absorbent elements. The material's ability for absorption, retention, and release of fluids or gases is directly influenced by its structural characteristics, including pore size, distribution, and connectedness. The activity of absorbent materials is included in many uses, namely wastewater purification and improved medical dressings are determined by the linked network of pores inside them. Various methods of characterizing have been explored, each providing distinct perspectives on the microstructure of absorbent materials. SEM offers researchers the ability to observe surface shape and pore structure with exceptional detail through higher clarity pictures. "Transmission Electron Microscopy (TEM)" allows for the examination of interior microstructures at the nanoscale, offering vital insights into the composition and crystallinity of materials. "X-ray Diffraction (XRD)" is extremely useful for investigating the crystallographic characteristics of materials that may absorb substances. This method facilitates the identification of the crystalline phases that are present and offers insights into the overall structure of the material. FTIR helps in comprehending the chemical composition of absorbent materials by identifying functional groups, which is essential for establishing a connection between microstructure and performance.

The significance of porosity in absorbent materials is of utmost relevance, and methods such as mercury intrusion porosimetry (MIP) and nitrogen adsorption play a critical role in assessing the volume and distribution of pores. Gaining comprehension of these factors facilitates the customization of materials for particular applications, guaranteeing the attainment of the highest level of performance. The Brunauer–Emmett–Teller (BET) technique is an effective way to calculate the surface area of absorbent materials, which is another important issue to consider. This knowledge is crucial for forecasting the material's ability to interact with and absorb liquids or gases. Through the cumulative utilization of these approaches, researchers gain a full comprehension of the complicated microstructure of absorbent materials.

The utilization of sophisticated methodologies such as AFM and Nuclear Magnetic Resonance (NMR) has yielded unparalleled understanding of the nanoscale characteristics and molecular behavior of absorbent materials. AFM enables researchers to investigate surface forces and interactions at the atomic scale, while NMR offers insights into molecule mobility, assisting in the development of materials with improved absorbent characteristics. The significance of microstructure characterization in absorbent materials is obvious in its multidisciplinary nature, affecting numerous sectors. Materials with well-specified microstructures are advantageous for environmental applications, such as water purification and pollution removal, since they enable effective absorption and retention. Absorbent materials are essential in the medical profession for wound care, medication administration, and implanted devices. Customizing microstructures enables the creation of materials with precise functions, catering to the distinct needs of each application. The progressive advancement of characterization techniques and the incorporation of computational approaches, such as molecular dynamics simulations, aid in achieving a more comprehensive understanding of the behavior of absorbent materials. The combination of experimental and computational methods enables the creation of materials that have improved performance, are cost-effective, and sustainable.

REFERENCES

1. Pellenz, L., de Oliveira, C. R. S., da Silva Júnior, A. H., da Silva, L. J. S., da Silva, L., de Souza, A. A. U., & da Silva, A. (2023). A comprehensive guide for characterization of adsorbent materials. *Separation and Purification Technology*, *305*, 122435.
2. Rio, S., Faur-Brasquet, C., Le Coq, L., & Le Cloirec, P. (2005). Production and characterization of adsorbent materials from an industrial waste. *Adsorption*, *11*, 793–798.

3. Ettish, M. N., El-Sayyad, G. S., Elsayed, M. A., & Abuzalat, O. (2021). Preparation and characterization of new adsorbent from cinnamon waste by physical activation for removal of Chlorpyrifos. *Environmental Challenges*, *5*, 100208.
4. Elizalde-González, M. P., Mattusch, J., Peláez-Cid, A. A., & Wennrich, R. (2007). Characterization of adsorbent materials prepared from avocado kernel seeds: Natural, activated and carbonized forms. *Journal of Analytical and Applied Pyrolysis*, *78*(1), 185–193.
5. Jin, Q., Dai, M., Zhan, X., Wang, S., & He, Z. (2022). Carbon nanotubes and graphene composites used in Cr(VI) detection techniques: A review. *Journal of Alloys and Compounds*, *922*, 166268.
6. Wu, H., & Vaneeckhaute, C. (2022). Nutrient recovery from wastewater: A review on the integrated Physicochemical technologies of ammonia stripping, adsorption and struvite precipitation. *Chemical Engineering Journal*, *433*, 133664.
7. Yang, R. T. (2003). *Adsorbents*. John Wiley & Sons. http://books.google.ie/books?id=7M3UkqQIaIEC&printsec=frontcover&dq=K.+
8. Amid, H., Mazé, B., Flickinger, M. C., & Pourdeyhimi, B. (2016). Hybrid adsorbent nonwoven structures: A review of current technologies. *Journal of Materials Science*, *51*(9), 4173–4200. https://doi.org/10.1007/s10853-016-9741-x
9. Dettmer, K., & Engewald, W. (2002). Adsorbent materials commonly used in air analysis for adsorptive enrichment and thermal desorption of volatile organic compounds. *Analytical and Bioanalytical Chemistry*, *373*(6), 490–500.
10. Demir, H., Mobedi, M., & Ülkü, S. (2009). Effects of porosity on heat and mass transfer in a granular adsorbent bed. *International Communications in Heat and Mass Transfer*, *36*(4), 372–377.
11. Mangwandi, C., Suhaimi, S. N., Liu, J. T., Dhenge, R. M., & Albadarin, A. B. (2016). Design, production and characterisation of granular adsorbent material for arsenic removal from contaminated wastewater. *Chemical Engineering Research and Design*, *110*, 70–81.
12. Swain, P. T. R., & Biswas, S. (2017). A comparative analysis of physico-mechanical, water absorption, and morphological behaviour of surface modified woven jute fiber composites. *Polymer Composites*, *39*(8), 2952–2960. https://doi.org/10.1002/pc.24294
13. Horoshenkov, K., Bécot, F. X., & Jaouen, L. (2010). Acoustics of porous materials: Recent advances relating to modelling, characterization and new materials. *Acta Acustica united with Acustica*, *96*(2), 5–6. https://doi.org/10.3813/aaa.918291
14. Zhang, J., Li, A., & Wang, A. (2006). Study on superabsorbent composite. VI. Preparation, characterization and swelling behaviors of starch phosphate-graft-acrylamide/attapulgite superabsorbent composite. *Carbohydrate Polymers*, *65*(2), 150–158. https://doi.org/10.1016/j.carbpol.2005.12.035
15. Ai, F., Yin, X., Hu, R., Ma, H., & Liu, W. (2021). Research into the super-absorbent polymers on agricultural water. *Agricultural Water Management*, *245*, 106513. https://doi.org/10.1016/j.agwat.2020.106513
16. Mehta, P., Sharma, M., & Devi, M. (2023). Hydrogels: An overview of its classifications, properties, and applications. *Journal of the Mechanical Behavior of Biomedical Materials*, *147*, 106145. https://doi.org/10.1016/j.jmbbm.2023.106145
17. Kóczán, Z., & Pásztory, Z. (2024). Overview of natural fiber-based packaging materials. *Journal of Natural Fibers*, *21*(1). https://doi.org/10.1080/15440478.2023.2301364
18. Han, Y., Quan, X., Zhao, H., Chen, S., & Zhao, Y. (2007). Kinetics of enhanced adsorption by polarization for organic pollutants on activated carbon fiber. *Frontiers of Environmental Science & Engineering in China*, *1*(1), 83–88. https://doi.org/10.1007/s11783-007-0016-2

19. Feng, Z. Y., & Meng, L. Y. (2021). Hierarchical porous carbons derived from corncob: Study on adsorption mechanism for gas and wastewater. *Carbon Letters*. https://doi.org/10.1007/s42823-021-00231-8
20. Wells, B. A., & Chaffee, A. L. (2011). Modeling gas separation in metal-organic frameworks. *Adsorption*, *17*(1), 255–264. https://doi.org/10.1007/s10450-010-9305-3
21. Qiao, Y., Zhan, Z., Yang, Y., Liu, M., Huang, Q., Tan, B., Ke, X., & Wu, C. (2021). Amine or Azo functionalized hypercrosslinked polymers for highly efficient CO2 capture and selective CO2 capture. *Materials Today Communications*, *27*, 102338. https://doi.org/10.1016/j.mtcomm.2021.102338
22. Hui Ying, N., & Yeong, Y. F. (2014). Heavy metal ions removal using amine-functionalized mesoporous adsorbents. *Applied Mechanics and Materials*, *625*, 65–68. https://doi.org/10.4028/www.scientific.net/amm.625.65
23. Knezevic, M., & Beyerlein, I. J. (2018). Multiscale modeling of microstructure-property relationships of polycrystalline metals during thermo-mechanical deformation. *Advanced Engineering Materials*, *20*(4). https://doi.org/10.1002/adem.201700956
24. Navrátilová, E., & Nedĕla, V. (2016). Characterization of burnt clays by X-ray diffraction analysis, chemical analysis and environmental scanning electron microscopy. *Microscopy and Microanalysis*, *22*(S3), 1862–1863. https://doi.org/10.1017/s1431927616010151
25. Falvo, M., Superfine, R., Washburn, S., Lord, S., Hall, A., & Jawerth, L. (2007). Nanoscale mechanical characterization using atomic force microscopy. *Microscopy and Microanalysis*, *13*(S02). https://doi.org/10.1017/s1431927607072017
26. Iwamiya, J., & Sinton, S. (1996). Stray-field magnetic resonance imaging of solid materials. *Solid State Nuclear Magnetic Resonance*, *6*(4), 333–345. https://doi.org/10.1016/0926-2040(95)01208-7
27. Chibani, S., & Coudert, F. X. (2020). Machine learning approaches for the prediction of materials properties. *APL Materials*, *8*(8), 080701. https://doi.org/10.1063/5.0018384
28. (2010). Advanced techniques for materials characterization (Adv. Mater. 10/2010). *Advanced Materials*, *22*(10). https://doi.org/10.1002/adma.201090030
29. Vranken, W. F. (2014). NMR structure validation in relation to dynamics and structure determination. *Progress in Nuclear Magnetic Resonance Spectroscopy*, *82*, 27–38. https://doi.org/10.1016/j.pnmrs.2014.08.001
30. Packwood, D., Nguyen, L. T. H., Cesana, P., Zhang, G., Staykov, A., Fukumoto, Y., & Nguyen, D. H. (2022). Machine learning in materials chemistry: An invitation. *Machine Learning With Applications*, *8*, 100265. https://doi.org/10.1016/j.mlwa.2022.100265
31. Liang, Z., Li, A., Deng, K., Ouyang, B., & Kan, E. (2023). Tailoring the microstructure of porous carbon spheres as high rate performance anodes for lithium-ion batteries. *Materials*, *16*(13), 4828. https://doi.org/10.3390/ma16134828
32. Silva-Alvarez, D., Dominguez-Lopez, I., Vidales Hurtado, M., Gutierrez-Antonio, C., Flores-Garay, K., & Garcia-Garcia, A. (2024). A review on the menagerie of green fluids and nanoparticles to develop sustainable biolubricant technologies. *Environmental Technology and Innovation*, *33*, 103532. https://doi.org/10.1016/j.eti.2024.103532
33. Goldstein, J. I., Newbury, D. E., Michael, J. R., Ritchie, N. W., Scott, J. H. J., & Joy, D. C. (2017). *Scanning electron microscopy and X-ray microanalysis*. Springer.
34. Eaton, P., & West, P. (2010). *Atomic force microscopy*. Oxford University Press.
35. Beckhoff, B., Kanngießer, B., Langhoff, N., Wedell, R., & Wolff, H. (Eds.). (2007). *Handbook of practical X-ray fluorescence analysis*. Springer Science & Business Media.
36. Delesma, F. A., Van den Bossche, M., Grönbeck, H., Calaminici, P., Köster, A. M., & Pettersson, L. G. (2018). A chemical view on X-ray photoelectron spectroscopy: The ESCA molecule and surface-to-bulk XPS shifts. *ChemPhysChem*, *19*(2), 169–174.

37. Mochizuki, S., & Zydney, A. L. (1993). Theoretical analysis of pore size distribution effects on membrane transport. *Journal of Membrane Science*, *82*(3), 211–227.
38. Abell, A. B., Willis, K. L., & Lange, D. A. (1999). Mercury intrusion porosimetry and image analysis of cement-based materials. *Journal of Colloid and Interface Science*, *211*(1), 39–44.
39. Yao, Y., & Liu, D. (2012). Comparison of low-field NMR and mercury intrusion porosimetry in characterizing pore size distributions of coals. *Fuel*, *95*, 152–158.
40. Myers, A. L., & Prausnitz, J. M. (1965). Thermodynamics of mixed-gas adsorption. *AICHE Journal*, *11*(1), 121–127.
41. Kiselev, A. V., & Yashin, Y. I. (2013). *Gas-adsorption chromatography*. Springer.
42. Dollimore, D., Spooner, P., & Turner, A. J. S. T. (1976). The BET method of analysis of gas adsorption data and its relevance to the calculation of surface areas. *Surface Technology*, *4*(2), 121–160.
43. Sing, K. S. (1989). The use of gas adsorption for the characterization of porous solids. *Colloids and Surfaces*, *38*(1), 113–124.
44. Chen, Y., Zou, C., Mastalerz, M., Hu, S., Gasaway, C., & Tao, X. (2015). Applications of micro-fourier transform infrared spectroscopy (FTIR) in the geological sciences a review. *International Journal of Molecular Sciences*, *16*(12), 30223–30250.
45. Zarnowiec, P., Lechowicz, L., Czerwonka, G., & Kaca, W. (2015). Fourier transform infrared spectroscopy (FTIR) as a tool for the identification and differentiation of pathogenic bacteria. *Current Medicinal Chemistry*, *22*(14), 1710–1718.
46. Andrade, J. D. (1985). X-ray photoelectron spectroscopy (XPS). *Surface and interfacial aspects of biomedical polymers: Volume 1 surface chemistry and physics* (pp. 105–195). Plenum Press.
47. Chastain, J., & King Jr, R. C. (1992). Handbook of X-ray photoelectron spectroscopy. *Perkin-Elmer Corporation*, *40*, 221.
48. Chen, W., Lu, F., Frew, D. J., & Forrestal, M. J. (2002). Dynamic compression testing of soft materials. *Journal of Applied Mechanics*, *69*(3), 214–223.
49. Talaat, A., Emad, A., Tarek, A., Masbouba, M., Essam, A., & Kohail, M. (2021). Factors affecting the results of concrete compression testing: A review. *Ain Shams Engineering Journal*, *12*(1), 205–221.
50. Prichard, J. C., & Hogg, P. J. (1990). The role of impact damage in post-impact compression testing. *Composites*, *21*(6), 503–511.

9 Advantages and Disadvantages of Novel Technologies for Adsorbent from Agro-Based Industries

B Yashvashri and Sumit Kumar

9.1 INTRODUCTION

Increasing demand for environmental remediation to be both sustainable and cost-effective has shone the spotlight on agro-industrial waste as a source for adsorbents. Industrial and agricultural activities have continued to grow with large quantities of by-products being generated, which, however, are mostly discarded as waste, contributing to environmental degradation. However, the latest trends in material science and green technologies have been focused on making a paradigm shift toward revalorizing these by-products into high-value materials, particularly for the adsorbents used in wastewater treatment, gas capture, and other environmental applications. Agro-industrial wastes like fruit peels, rice husks, coconut shells, and sugarcane bagasse are sufficiently abundant in lignocellulosic components, which are very promising in terms of adsorption properties after proper physical and chemical modification. These cheap materials will not only reduce waste disposal problems but also offer alternative, renewable sources of adsorbents instead of activated carbon. New technologies, including thermal treatments, surface functionalizing, and nanotechnology, are increasingly applied to develop competitive agro-waste-based adsorbents by achieving better adsorption capacity, higher selectivity, and better potential renewability. Despite the positive trends, the use of agro-industrial waste as the basis for adsorbent production entails inherent challenges and limitations. Some of these include variability in raw material composition, some energy-intensive processes during modification, and the leaching of undesirable substances. The remaining issues that pose significant critical questions with regard to economic scalability, as well as industrial adoption, will also require further investigation. In this chapter, the advantages and disadvantages of novel technologies for converting agro-industrial waste to an adsorbent with effective properties are explored. Special attention is given to comparative performance with traditional materials, the environmental and

 DOI: 10.1201/9781003767251-9

economic implications of their production, and directions in research and application in the future. By this, the chapter aims to present a comprehensive understanding of the current status and potential of agro-industrial wastes as a sustainable source for adsorbent technologies in the global context.

9.2 NOVEL TECHNOLOGIES USED IN ADSORBENTS

Recently, agro-industrial waste has become a promising area of application for adsorbents in water treatment due to its cost-efficiency and environmental advantages. Several innovative technologies have been developed to enhance the performance of these adsorbents, which can be used as an alternate material instead of a conventional one. Agro-based materials can thus be used in their raw form or modified through processes such as carbonization to enhance the adsorption capacity (Aftab et al., 2024). Sugarcane bagasse was also converted into activated carbon, which achieved remarkable efficiency in the removal of pharmaceuticals and personal care products (Sethi, 2024). Nevertheless, such activated carbons suffer from some drawbacks, aside from the previously mentioned issues of generation and scalability (Rangappa et al., 2024).

9.2.1 Pyrolysis

Agro-industrial waste production into adsorbents through pyrolysis has multifaceted advantages that make the environment much more sustainable. It not only transforms waste into useful materials but also facilitates waste management and resource recovery (Chouikhi et al., 2021).

9.2.1.1 Advantages

Pyrolysis of agro-industrial waste is a sustainable route that can optimize the properties of carbon materials as part of strategies for waste management and carbon capture. Using different pyrolytic temperatures for synthesis, biochars were produced, while activated carbons were obtained from biochar that received variable dosages of activating agents at the same pyrolytic temperature of 500°C. Biochars were prepared under increasing pyrolytic temperatures, with generally higher adsorption capacities, and were stable throughout multiple cycles when prepared at the higher temperature range. Chemical activation of biochar using potassium hydroxide (KOH) significantly increased its microporosity and CO_2 adsorption ability than some samples that have 3.6 mmol of CO_2 per gram at 25°C and 760 mmHg. However, for N_2 and H_2, its adsorption capabilities were much lower; hence, these biochars and activated carbons were highly selective in the capture of CO_2 (Chouikhi et al., 2021).

The biochars derived from other biomass sources can possess surface areas up to 450 m²/g, which indeed plays an important role in the effective adsorption of contaminants by using biochar. Oxygen-containing functional groups further enhance their binding abilities for heavy metals. For example, it has been established that post-treated biochars significantly enhance the adsorption capacities for lead (Pb^{2+}) and copper (Cu^{2+}). Furthermore, mechanical and chemical treatments, such as sodium hydroxide (NaOH) treatment, have also been shown to exhibit enhancements in the performance of the adsorption capacity of the biochar by increasing its surface area as well as incorporating the functional groups (Cai et al., 2021). Biochar is very effective in soil remediation, water treatment, and air purification because of its advantageous structural and surface characteristics (Illankoon et al., 2023). Particularly at higher pyrolysis temperatures, biochars made from lignocellulosic feedstocks like sawdust and wheat straw show enhanced aromaticity and greater carbon stability (Nzediegwu et al., 2021).

9.2.1.2 Disadvantage

The development of pyrolysis has shown great promise for improving the performance of adsorbents derived from agricultural sources. Still, a number of challenges need to be overcome. One significant problem is the variability in biochar's adsorption capacity, which is impacted by a number of variables including the kind of feedstock utilized and the particulars of the pyrolysis process. For example, biochar's aromaticity is enhanced and its polarity is decreased at higher pyrolysis temperatures. Its ability to effectively adsorb non-polar organic compounds is adversely affected by this shift; this subject is covered in more detail in the sections that follow. Studies carried out by B. Chen et al. (2008) found that the high temperatures used in pyrolysis cause a decrease in polarity and an increase in aromaticity, which decreases the ability of biochar to adsorb non-polar organic molecules. Pyrolysis by-products such as tar and volatile organic compounds (VOCs) should be handled carefully because of the serious potential environmental effects. During pyrolysis, CO_2 promotes the thermal cracking of VOCs, which lowers tar formation and increases syngas production. Numerous biomass types, such as peat, agricultural wastes, and oak tree residues, have been shown to exhibit this phenomenon (Lee et al., 2017).

Biochar properties are mostly determined by the choice of feedstock; nevertheless, wood-based biochar typically has the largest surface area, while crop- and grass-based biochars have the highest cation exchange capacities (Ippolito et al., 2020). According to Hassan et al. (2020), biochars made from hardwood and softwood have greater surface area and carbon content than biochars made from manure and grass, but they also have lower concentrations of oxygen and mineral components. In spite of its potential energy efficiency, pyrolysis, for instance, nevertheless uses a lot of energy unless energy recovery mechanisms are included. Methods like co-pyrolysis with additives, integrated processes, and catalytic pyrolysis improve the amount of energy recovered while lowering the emissions related to the process. If not properly handled, incomplete pyrolysis of lignocellulosic mass may also lead to the production of dangerous gases including CO_2 and methane, increasing greenhouse gas emissions (Rangel et al., 2023). Hence, issues such as energy consumption and adsorption variability must be resolved. To maximize the advantages of biochar in pollutant removal applications, more research is essential to optimize feedstock selection and pyrolysis conditions.

9.2.2 Hydrothermal Carbonization

Hydrothermal carbonization (HTC) is a promising process for the production of hydrochar from biomass as an effective adsorbent. HTC's ability to operate at low temperatures (between 180 and 250°C) and high pressures makes it suitable for direct use of high-moisture biomass without the need for expensive pre-drying processes. Because of this flexibility in process conditions, hydrochar's properties can be tailored to improve its adsorption capabilities for pollutants such as organic compounds and heavy metals (A. Jain et al., 2014).

9.2.2.1 Advantages

HTC efficiently converts biomass into high-surface-area hydrochar, suitable for adsorbing pollutants with minimal energy consumption and greenhouse gas emissions involved during adsorption (Alatalo & Sillanpää, 2020). This process effectively makes use of agro-industrial waste and avoids associated disposal problems, bringing out environmentally friendly alternatives to synthetic adsorbents. The study proves that orange peel waste can be well converted into hydrochar through HTC with high carbon content and adjustable properties that can be tuned for specific adsorption applications and high carbon content (Satira et al., 2021). Based on these characteristics of hydrochars, the operating conditions, such as temperature, pressure, and time, may be changed to vary surface areas and functional groups to optimize hydrochars for targeted contaminants. Hydrochars obtained at different temperatures were analyzed for their physicochemical and combustion properties. The hydrochar mass yield decreased with an increase in temperature from 72.45% at 150°C to 41.88% at 300°C, whereas the higher heating value increased from 19.22 MJ/kg at 150°C to 29.97 MJ/kg at 300°C, which indicates the potential of higher-temperature hydrochars being used for energy-dense fuel applications, in particular with bioenergy production (C. Chen et al., 2021). Additionally, the atomic ratios of hydrogen to carbon and oxygen to carbon decreased from 1.30 and 0.66 at 150°C to 0.77 and 0.27 at 300°C, respectively. This reduction in atomic ratios implies greater carbon stability, which is beneficial for long-term carbon sequestration in soil, helping in climate change mitigation efforts (C. Chen et al., 2021).

The adsorption capacity of hydrochars was discovered to be improved because their surface structure was rougher and various pore structures were seen utilizing SEM examination at 240–300°C. This makes it perfect for environmental uses like industrial pollutant removal or water purification. Waste *Camellia oleifera* shells had different combustion behaviors, and their HTC; the stability of hydrochars depended upon the rising in temperature of HTC (C. Chen et al., 2021). HTC works in water temperatures of 180–300°C with far lower inputs of energy than traditional pyrolysis and is very efficient in the process converting biomass into hydrochar, stabilizes the carbon reservoir, decreases the emission of greenhouse gases, and produces adsorbents of enormous surface areas for pollutant adsorption (Czerwińska et al., 2022). Hydrochar is generally characterized with a high surface area and porosity, and hence post-treatment, it further enhances (Fiori et al., 2014). Such a parameter explains why hydrochar proves to be particularly useful in removing contaminants such as dyes and heavy metals from water, making it an eco-friendly method of environmental restoration (Czerwińska et al., 2022).

9.2.2.2 Disadvantages

Wet biomass can be converted into useful products using the HTC process; however, there are a number of drawbacks, including infrastructure costs, energy consumption, and process control. HTC operates in environments with high temperatures and pressure, necessitating the use of specialized equipment that comes with a high initial cost and ongoing maintenance expenses. When compared to more conventional techniques like pyrolysis, these costs are typically higher (Díaz-Rasero et al., 2024). Energy recovery systems can reduce some operating expenses, but adding them raises HTC's overall operating costs even more (Shah et al., 2024). HTC requires a significant energy input to maintain the required temperature and pressure conditions throughout the process, even though it does not require biomass to be pre-dried (Devnath et al., 2024).The performance variation of HTC depending on feedstock is one of its main drawbacks. Depending on the kind of biomass utilized, HTC's efficiency at pilot scales can vary from 26% to 63%, which adds unpredictability (Shah et al., 2024). Therefore, achieving these crucial parameters, particularly pressure and temperature, requires careful management.

Maintaining control over these crucial variables, particularly pressure and temperature, is necessary to provide consistently high-quality products. Variations in these parameters may result in changes to the hydrochar's surface characteristics and adsorption efficiency (Alonso et al., 2024). In general, hydrochar has less surface area and porosity than activated carbon, which reduces its effectiveness as an adsorbent. While activation processes can improve these qualities, a lot of traditional activation techniques release pollutants into the environment (Sun et al., 2016). Furthermore, HTC's liquid effluents frequently contain organic compounds that need to be carefully managed to avoid contaminating the environment. Additionally, hydrochars might contain polycyclic aromatic hydrocarbons (PAHs) and heavy metals, which are cumulative and dangerous substances that call for careful examination and sensible waste management techniques (Babeker & Chen, 2021).

9.2.3 Surface Functionalization of Adsorbents

Surface functionalization of adsorbents derived from agro-based industries enhances their efficiency in wastewater treatment by modifying their chemical properties. This process typically involves the application of surface treatment agents that react with the adsorbent particles, often facilitated by hydrothermal methods to create functionalized surfaces capable of selectively binding contaminants.The functionalization can significantly improve adsorption capacities, making these modified materials comparable to commercial adsorbents like activated carbon (N. Jain et al., 2008). Techniques such as controlling the spacing of functional groups and selective site functionalization are crucial for optimizing the adsorption characteristics of these agro-based materials (Yoshitake, 2010). Additionally, the performance of these adsorbents is influenced by factors such as surface area, pH, and the ionic potential of the solution, which are critical for effective metal ion removal from wastewater (Emenike et al., 2016). While the benefits of surface functionalization are clear, challenges remain in optimizing these processes for broader commercial applications, including cost-effectiveness and scalability.

9.2.3.1 Advantages

Surface functionalization of adsorbents from agro-waste greatly enhances the level of capture of different types of pollutants, such as heavy metals and organic contaminants. The application of suitable functional groups such as carboxyl and hydroxyl allows to design the adsorbents with selectivity toward particular adsorption; hence, this improves efficiency in treating wastewater. Functionalization increases the number of active sites around the surface of the adsorbent, contributing to increase the adsorption capacity. For instance, functionalized agricultural wastes show higher adsorption of heavy metals than unfunctionalized counterparts. Moreover, one attains better reusability since the functionalization of the surface allows easier regeneration when removing the pollutant which enhances lifespans. It, therefore, allows easy, cost-effective, and sustainable substitutability from synthetic-based adsorbents with minimal environmental footprints. The agro-waste materials in this context is employed toward application in the management of waste to facilitate environmentally friendly practices from such a perspective (Ogunlalu et al., 2021).

9.2.3.2 Disadvantages

Surface functionalization of agro-waste materials poses many significant challenges that include high costs involved, risks associated with environmental impact, and potentially declining adsorption capacity. Hazardous chemicals applied in the extensive process of functionalization lead to severe environmental contamination and health risks (García, 2011).

Functionalization often uses expensive chemicals, which outweigh the benefits of using agro-waste as a base material. The multi-stage processes are complicated because of high labor and energy costs; therefore, it is difficult to produce in large quantities (Yoshitake, 2010). Some of these chemicals are harmful to the environment and humans; hence, they need to be handled or disposed with care (García, 2011). Poor control leads to extreme pollution to the environment. Bulk functional groups may occupy active sites on the adsorbent, causing it to reduce the overall adsorption capacity (Goor et al., 2017). Uniform functionalization is also challenging to achieve, resulting in batch-to-batch variation in performance (Yoshitake, 2010).

9.2.4 Nanotechnology

The integration of nanotechnology in the design of adsorbents from agro-based industries enhances various advantages that improve the efficiency, selectivity, and functionality.

9.2.4.1 Advantages

Agro-waste, used as a base material for nano-material, promotes sustainability and reduces the dependency on synthetic adsorbents (Aguilar-Pérez et al., 2022). Nanomaterials display some unique properties that elevate the capabilities of an adsorbent for environmental applications. They are mostly characterized by a high surface-to-volume ratio, which increases active sites that have the potential to interact with pollutants, thereby enhancing efficiency in capturing contaminants with low concentrations. This is very beneficial in water purification and air filtration applications.

The functionalized surfaces of nanomaterials, such as metal oxides and graphene oxide, offer specific chemical sites for the binding of pollutants, making them excellent at adsorbing organic pollutants and heavy metals (Siddeeg et al., 2020). Their nanoscale size also allows for faster adsorption rates by shortening the diffusion paths between the pollutant and active sites (Sarkar et al., 2021). For environmental clean-up applications, nanomaterials are very effective because of their small particle size and surface functionality, which leads to faster and more efficient pollutant removal.

Nanotechnology provides control over surface properties as specific functional groups can be attached to target pollutants such as dyes and pharmaceuticals (Agboola et al., 2021). Such adjustment enhances selectivity and efficiency in the removal of contaminants (Siddeeg et al., 2020). Nano-adsorbents are economical for long-term use in pollutant removal because they frequently show good reusability and retain their adsorption capacity over several cycles. Their robust structural integrity and the reversible adsorption process, which makes it simple for contaminants to be desorbed and free up the active sites for additional adsorption, are credited with their longevity. Nano-adsorbents' capacity for regeneration and reuse improves sustainability by prolonging their useful life and lowering the need for constant material replacement (Siddeeg et al., 2020).

9.2.4.2 Disadvantages

Although the integration of nanotechnology in the creation of adsorbents for agro-based industries offers creative approaches to waste management and pollutant removal, there are a number of important obstacles to overcome. One significant concern is the possible toxicity of nanoparticles, which may pose risks to ecosystems and human health because of their small size and strong reactivity. These substances have the potential to accumulate over time and cause long-term health and ecological issues. Furthermore, releasing nanomaterials into the environment may have unanticipated ecological repercussions that degrade the quality of the soil and water (Wahab et al., 2024). The high cost of nanomaterial synthesis and deployment limits their economic viability and widespread adoption in agro-based sectors. Furthermore, the absence of thorough laws regulating nanotechnology in agriculture makes it more difficult to apply it safely and effectively (Ullah et al., 2024). The complex procedures needed to produce useful nanomaterials from agricultural waste may not be accessible in all areas due to the need for cutting-edge technologies (Mandal et al., 2024). These difficulties highlight the necessity of a well-rounded strategy that takes into account both the possible advantages of nanotechnology in improving agricultural methods and environmental sustainability, as well as the risks connected with its use. The ethical and sustainable application of nanotechnology in agro-industries depends on resolving safety issues, reducing negative effects on the environment, and enhancing economic viability.

9.2.5 Microwave-Assisted Synthesis

Microwave-assisted synthesis has emerged as a promising method for producing adsorbents from agro-industrial waste, enhancing their efficiency in environmental applications. This technique not only accelerates the synthesis process but also improves the properties of the resulting materials. While microwave-assisted methods significantly improve the properties of adsorbents derived from agro-industrial waste, challenges remain in optimizing these processes for large-scale applications and ensuring the sustainability of the materials used.

9.2.5.1 Advantages

Microwave-assisted synthesis has significant advantages over the conventional thermal methods in the production of adsorbents. This technique enables rapid heating and selective energy application, and thus the reaction kinetics are enhanced along with improved quality of the product obtained. Compared to traditional methods, this procedure greatly increases the yield and shortens the time required for such reactions to occur (Baghurst & Mingos, 1988). Microwaves cause volumetric heating, allowing materials to be heated uniformly from within, which reduces energy waste and minimizes localized overheating (Schwenke et al., 2015). Controlled heating results in the development of excellent surface area and pore structure for adsorbents, which are essential for efficient pollutant capture (Polshettiwar et al., 2009). Microwave synthesis also results in an increased number of surface functional groups, leading to improved adsorption capacity with a variety of contaminants (Rao & Ramesh, 1995). In line with sustainability objectives, using agro-waste and green solvents in microwave synthesis lowers energy consumption and chemical waste (Polshettiwar et al., 2009; Rao & Ramesh, 1995).

9.2.5.2 Disadvantages

Although microwave-assisted synthesis has many benefits, such as quick processing and energy savings, it also has significant drawbacks that prevent large-scale use. Special microwave reactors are very expensive, thus increasing the capital investments for industries at the initial stages. Operating costs concerning the maintenance of such reactors can also be high, hence affecting economic viability throughout the stages. Scalability issues with microwave-assisted synthesis arise because larger reactors may result in uneven microwave distribution, which can cause uneven heating and variable product quality (Cerón-Camacho, 2024). Overheating in hot spots, mainly for heat-sensitive agro-based materials, may lead to degradation of their properties and reduce adsorption efficiencies. Although microwave synthesis is faster than conventional methods, it might eventually end up consuming massive amounts of energy, which degrades its overall efficiency. Finding optimal microwave parameters requires extensive trial and error, which further complicates the process for industries (Pandhurnekar et al., 2024). On the other side, many researchers think that individual benefits achieved by microwave processing, like lower production time and properties of materials, will later outweigh these challenges to increase the industries' acceptance (Pandhurnekar et al., 2024; Roy et al., 2023).

9.2.6 Ionic Liquid-Based Processing

Processing agro-industrial waste using ionic liquids (ILs) to create adsorbents offers a promising approach to heavy metal ion remediation. This method makes use of the special qualities of ILs to improve agricultural waste materials' adsorption capacities, making them competitive substitutes for traditional adsorbents. Although IL processing exhibits a lot of promise, there are still issues with maximizing the method's scalability and economic feasibility for broad industrial use.

9.2.6.1 Advantages

The ILs are considered green solvents since they are non-volatile and possess low vapor pressure, which greatly contributes to reducing environmental pollution and occupational hazards. Its unique properties enable the extraction of valuable constituents from the agro-waste to enhance the performance of adsorbents in environmental applications. When ILs are used in place of volatile organic solvents (VOS) in a variety of industrial processes, air pollution is effectively reduced (Valli Sowbhagyam, 2024). When compared to conventional VOS, their low flammability and minimal vapor pressure provide far safer substitutes (Gujjala et al., 2024). Furthermore, cellulose, lignin, and hemicellulose—essential components in building porous structures that increase adsorption capacity—can be extracted more easily with the help of ILs (Z. Wang et al., 2023).

ILs also have the benefit of improving adsorbents interaction with pollutants like dyes and heavy metals by changing their surface functional groups (P. Wang et al., 2024). This potential can be optimized for processing various kinds of agro-waste by modifying the cation–anion pairing in ILs (Gujjala et al., 2024). In addition, a large number of ILs are recyclable, which lowers the overall cost of manufacturing high-performance adsorbents while also strengthening sustainability (P. Wang et al., 2024). Despite their benefits, ILs may still pose environmental risks, especially when it comes to their biodegradability and long-term effects. Future studies are required to completely comprehend these possible drawbacks and guarantee their safe use in environmental and industrial processes (Z. Wang et al., 2023).

9.2.6.2 Disadvantage

Agro-based industries face both opportunities and challenges with ILs, especially when it comes to their environmental impact and economic viability. ILs have the potential to be environmentally friendly solvents, but their high synthesis costs, difficult recycling procedures, and potential toxicity make them unsuitable for widespread use. ILs can be prohibitively expensive to synthesize and purchase, which restricts their use in large-scale manufacturing (P. Wang et al., 2024). Thus, ILs increase operating costs because they need specific knowledge to handle safely (Z. Wang et al., 2023). Though ILs are frequently thought of as environmentally benign, some varieties may be harmful to soil microbes, which can have an impact on enzyme activity and biodiversity. Aquatic ecosystems and soil health may be at risk from improper disposal or unintentional release (Z. Wang et al., 2023). Additionally, compared to conventional solvents, ILs frequently have higher viscosities, which impair mass transfer and slow reaction rates during the preparation of adsorbents (Valli Sowbhagyam, 2024). The intricacy of recovering and purifying ILs for reuse can result in energy-intensive procedures, making their wider use even more challenging (Kumar, 2024). To balance economic and environmental concerns, however, further research into optimizing ILs for particular agro-waste materials may lead to creative solutions.

9.2.7 Biological and Enzymatic Treatments

An environmentally friendly method of wastewater remediation is offered by biological and enzymatic processes that use agro-industrial waste as adsorbents. These techniques increase the efficiency of agricultural by-products in eliminating pollutants by utilizing their natural qualities. Although agro-industrial waste has a lot of potential as an adsorbent, there are still issues with optimizing these materials for different wastewater compositions and guaranteeing consistent performance in a range of environmental conditions.

9.2.7.1 Advantages

Biological and enzymatic treatments have been a substitute for agro-industries that are environmentally friendly as compared to the chemical methods. These treatments utilize natural phenomena to minimize impacts on the environment and maximize the worth of agro-waste and ensure safe and effective delivery. The enzymatic process produces fewer toxic products as compared to the chemical treatment. This implies that there is a lesser hazard that exists to the environment (Singh et al., 2020). Biological treatments usually occur at relatively mild conditions, which leads directly to lower energy consumption with a smaller carbon footprint, according to Bhardwaj et al. (2021). In this case of converting low-value agro-waste into high-value adsorbents, it offers not only waste management but also economic returns, in the conversion process indicated by Alcántara et al. (2022). The biological adsorbents also tend to be biodegradable, which reduces the waste disposal problems associated with the conventional materials (Alcántara, 2019). The "marketability" of adsorbents against contaminants was enhanced through improving adsorption characteristics from biological treatments, according to Osuoha & Nwaichi (2021). Although enzymatic and biological treatments have several merits, scaling up these processes for large-scale industrial application is still an issue and thus more research and development work is required (Singh et al., 2020).

9.2.7.2 Disadvantages

Use of enzymes in biological treatment processes is plagued with several problems that prevent these systems from being economical and effective at the site scale. These include the high production cost, elaborate procedure, and sensitivity of the process to environmental conditions. The enzymes used in the processes are expensive to manufacture or obtain, heavily increasing the cost of production of the adsorbent (Singhania et al., 2015). Production costs are increased by the need to continuously replenish enzymes because they deactivate under harsh conditions. (Bhardwaj et al., 2021). Biological processes in fermentation and enzyme reactions take several hours to days, making this slower compared to the other physical and chemical methods (Vermelho et al., 2012). The enzymes are sensitive to variations in pH, temperature, and substrate concentration. Such variations make it very difficult to control these reactions precisely that in turn affects the quality of the product (Kirk et al., 2004). Sterile conditions in preventing contamination tend to be expensive and rather challenging to attain. Despite all these challenges, however, trends in protein engineering and enzyme technology are taking the field toward more economical and efficient applications in industries (Bhardwaj et al., 2021).

9.2.8 3D Printing and Additive Manufacturing

In the field of environmental remediation, 3D printing and additive manufacturing have become revolutionary technologies for producing adsorbents from agro-industrial waste. This creative method increases the efficacy and efficiency of adsorbents while also making use of inexpensive materials. Although 3D printing has great potential for using agro-industrial waste to create adsorbents, overcoming technical obstacles is necessary to fully reap the benefits of this technology in environmental applications.

9.2.8.1 Advantages

3D-printing technology significantly outperforms the design of adsorbents and their production by means of generated geometries that maximize adsorption properties. This process does not waste materials and encourages the use of agro-based inputs, aligning with sustainable practices. The capability of rapid prototyping through 3D printing facilitates fast iteration cycles that lead to effective adsorbents in a very short time. 3D printing enables complex structures by increasing the surface area and pore distribution, which in turn increases the efficiency of the adsorption process. Wastewater purification applications demonstrate the superior catalytic and adsorption capabilities of hierarchically porous materials, like those created through 3D printing (C. Cai et al., 2021). Utilizing agricultural waste in the 3D-printing process encourages the use of renewable resources and reduces the usage of synthetic materials. The lower cost of set-up related to the three-dimensional printing makes it suitable for small-scale production of adsorbents, allowing for experimental applications.

Recent developments in additive manufacturing will make it possible to develop larger scales, and this technology can transform industrial practice in the manufacture of adsorbents. Despite the advantages offered by three-dimensional printing, there are challenges in scaling up these technologies as widely used industrial applications because of inconsistencies in their performance from one larger batch to another.

9.2.8.2 Disadvantages

The use of 3D printing in the preparation of adsorbents on agro-based materials apparently has a lot of drawbacks when considering effectiveness and applicability. In comparison to conventional techniques, the mechanical qualities and surface features of printed adsorbents frequently fall short, despite the fact that 3D printing opens up new design possibilities. Not all agro-based materials are endowed with properties suitable for effective 3D printing, which may reduce design options. 3D-printed adsorbents suffer from poor mechanical strength and stability, reducing the performance in dynamic situations. The optimization of printing parameters like temperature and speed is not easy and requires massive experimentation. Surface roughness due to 3D printing reduces the actual area for adsorption, thus degrading the performance in general. Specific steps like curing or sintering are required for enhancements in material properties, which may complicate the production workflow. Initial equipment costs, including the cost of a good printer and materials, entail an initial investment, hence not economical to make in large quantities compared to other methods. Though this method has various drawbacks, constant studies on material systems together with their techniques are seeking ways of enhancing strength and stability for 3D-printed adsorbents to overcome some disadvantages.

9.2.9 Electrospinning and Nanofiber Synthesis

A promising method for creating nanofibers (NFs) from agro-industrial waste is electrospinning, which provides a sustainable way to create efficient adsorbents for environmental remediation. This technique increases the NFs' capacity to adsorb pollutants by taking advantage of their large surface area and porosity. Although electrospinning presents novel approaches to waste valorization and pollutant removal, there are still difficulties in tailoring the characteristics of NFs for particular uses, such as stability and reusability in diverse environmental settings.

9.2.9.1 Advantages

The high surface area-to-volume ratio of electrospun NFs renders them exceptionally efficient in the adsorption of pollutants. By giving precise control over their properties, this process enhances the use of fibers in environmental remediation. The flexibility of electrospinning permits the use of biodegradable materials and, in turn, promotes sustainability in the adsorbent production process. Due to their high porosity and abundance of adsorption sites, electrospun polymeric NFs can effectively remove heavy metals and dyes from water.

Functionalization of the surface during or after electrospinning can have a more positive impact on adsorption, thus targeting harmful pollutants. Electrospinning enables the production of fibers with controlled diameters and orientations, resulting in enhanced adsorption efficiency by optimizing the flow pathways. This influence can also result in better performance in applications such as air filtration and water purification. Using renewable agro-based materials in electrospinning aligns with eco-friendly procedures and helps create green adsorbents. The technique is also affordable for small-scale production because it requires less involvement of extensive machinery when compared to traditional methods. Although electrospun NFs can break easily, reinforcement techniques enhance their strength without compromising their potential for adsorption or efficiency, making these NFs multifunctional across various industries.

9.2.9.2 Disadvantages

Electrospinning is an extremely intricate process that also poses several challenges regarding industrial scalability and material versatility. Some of the vital operational parameters that include voltage, flow rate, and the distance between the needle and collector significantly affect the nature of fibers. Due to these factors, optimization of this process becomes difficult. Certain agro-based materials cannot be electrospun successfully because of restrictions on viscosity and fiber formation. An agro-based material's low viscosity causes the solution to become excessively fluid, resulting in bead formation rather than smooth fibers. On the other hand, excessive viscosity can cause the solution to flow improperly and completely inhibit the formation of fibers.

The low production rate of NFs in comparison to conventional methods poses a challenge for large-scale applications.

Electrospinning is energy-intensive, particularly in supporting high-voltage levels, which may result in higher operating costs. The requirement for stabilization as a post-processing step makes the production process complicated and time-consuming. Electrospun fibers possess weaker mechanical properties than conventional materials. This makes electrospinning unsuitable or less efficient in various applications. Strict safety precautions are required due to the health risks associated with the use of VOS. In spite of the obstacles, ongoing studies are striving to improve electrospinning methods and broaden the scope of compatible materials, which could boost its industrial relevance and applicability.

9.2.10 Ultrasound

Using high-frequency sound waves, usually above 20 kHz, ultrasonic technology produces a variety of physical and chemical effects on materials. The use of ultrasound in materials science has increased in recent years, particularly to improve adsorbents produced from agro-industrial wastes. Ultrasound's cavitation effect, when combined with improved chemical reactions and rapid mass transfer, makes it a potentially helpful method for enhancing the characteristics and performance of adsorbents in various environments.

9.2.10.1 Advantages

Ultrasound significantly enhances mass transfer rates during adsorption processes through acoustic cavitation, which generates microstreams that improve particle dispersion and contact between adsorbents and adsorbates. Increasing the porosity and surface area of adsorbents speeds up the adsorption process and improves the adsorption capabilities for various contaminants. The collapse of bubbles during cavitation creates localized high temperatures and pressures, facilitating chemical reactions and improving mass transfer coefficients. Ultrasound reduces the activation energy for surface diffusion, enhancing desorption rates and overall adsorption efficiency. Ultrasound-assisted processes often require fewer chemicals, aligning with sustainability goals in adsorbent production. These processes can be conducted at lower temperatures, reducing energy consumption and costs. Ultrasound can be applied to various agro-based materials, allowing for the development of diverse adsorbents. The technology is suitable for large-scale applications, meeting industrial demands while maintaining efficiency. Although ultrasound has many benefits, challenges such as high cost of the equipment and requirement for application-specific optimization remain.

9.2.10.2 Disadvantages

There are advantages and disadvantages to using ultrasonic technology in agro-based applications. Despite its potential to improve product quality and shorten processing times, a number of obstacles may prevent its widespread adoption, especially for small-scale businesses. Ultrasound equipment can be expensive to buy initially, which may discourage small businesses and start-ups from implementing this technology. While ultrasonography can shorten processing times, its environmentally favorable advantages may be outweighed by the high energy costs associated with maintaining ultrasonic waves.

Ultrasound may not always work well with agro-based materials; some may deteriorate or lose their desired qualities when exposed to high-frequency sound waves. The efficacy of adsorbents may be adversely affected by the breakdown of organic molecules caused by prolonged exposure. Optimizing process parameters through effective ultrasound application necessitates specific knowledge, which could be a disadvantage for facilities with less experience. It is difficult to prepare adsorbents consistently because of things like sample heterogeneity and differences in ultrasonic intensity. Despite these challenges, the potential advantages of ultrasonic technology in improving product quality and efficiency call for more research and development in agro-based applications.

9.2.11 Plasma Technology

Plasma is considered an inventive method of material science through which synthesis and modification of adsorbents occur. Plasma is a fourth state of matter, consisting of dissociated gases with free electrons, ions, and neutral particles. Plasma is created under different conditions, resulting in different plasma types: thermal, non-thermal, low-pressure, and atmospheric pressure plasma, which are assigned unique properties and applications. The plasma technology seems to be a developing method for the sustainable transformation of organic materials into high-performance adsorbents toward the clean-up of the environment.

9.2.11.1 Advantages

Plasma treatment is a revolutionary way of improving the properties of adsorbents to effectively capture pollutants. This process works at lower temperatures and gets rid of hazardous solvents, which not only increases adsorption capacity but also supports sustainable practices. Plasma treatment increases the surface area and introduces functional groups, which enhance contaminant adsorption. For example, activated carbon fibers treated with plasma demonstrated a 20.9% increase in Acid Orange II adsorption capacity. Chromium uptake in natural zeolites treated with plasma increased by 130% the effectiveness of this technology to improve adsorption performance. Since plasma techniques do not require solvents and operate at room temperature, they are environmentally friendly and consistent with sustainable material production methods. Different adsorbents can be produced by applying plasma technology to a range of agro-based materials. This flexibility is essential for satisfying industrial demands. While plasma technology shows considerable potential, issues remain in optimizing treatment settings and guaranteeing consistent performance across different materials. To completely comprehend the long-term stability and financial feasibility of these therapies in industrial applications, more investigation is required.

9.2.11.2 Disadvantages

Small-scale operations in agriculture and other industries face major obstacles due to the high prices and complexity of plasma technology. Key challenges include the costly equipment, the need for exact control over treatment parameters, and the possibility of material degradation.

The high technological cost of plasma treatment equipment may discourage new competitors. Managing parameters such as gas composition and pressure requires specialized knowledge due to operational complexity, which makes optimization attempts more difficult. The spectrum of useful materials is limited since not all agro-based materials can survive plasma treatment; some may deteriorate. Adsorbents may sustain damage from improperly managed plasma processes, which could impair their functionality and mechanical strength. Strict safety precautions are required because plasma technology can produce dangerous by-products. Although plasma processes are rapid, they can use a lot of energy, which could negate any environmental advantages). Despite these obstacles, research into plasma applications is still being conducted to find creative solutions, suggesting that this sector may see further developments in the future.

9.3 CONCLUSION

This chapter investigates how adsorbents made from agro-industrial wastes might be used to address today's environmental problems. There are several advantages to repurposing agricultural waste for this purpose, including economic savings, sustainability support, and waste management. Modern methods, including thermal processing, chemical activation, and nanotechnology, have increased these materials' specificity and efficiency, making them viable substitutes for conventional activated carbon. However, a number of obstacles prevent these adsorbents from being widely used. Notable technical obstacles include variations in the composition of raw agro-wastes, the high energy and resource requirements of modification techniques, potential leaching problems, and the instability of the materials. Other challenges include the economic feasibility of large-scale manufacturing and practical applications.

In order to progress this area, future studies must concentrate on improving production techniques to produce high-performance adsorbents using integrative approaches that strike a compromise between sustainability and efficacy. Cost–benefit analysis and life cycle assessments will be essential for comprehending how new technologies will affect the environment and the economy. In summary, agro-industrial waste-based adsorbents have a lot of potential to help reduce environmental issues, but it is important to examine both their advantages and disadvantages. Researchers, companies, and legislators must collaborate to progress this sector and promote the sustainable use of these materials for environmental cleaning.

REFERENCES

Aftab, R. A., Yusuf, M., Ahmad, F., Danish, M., Zaidi, S., Vo, D. V. N., Nguyen, A.-T., Rahman, M. M., & Ibrahim, H. (2024). Green technology approach towards the removal of heavy metals, dyes, and phenols from water using agro-based adsorbents: A review. *Chemistry – An Asian Journal*. https://doi.org/10.1002/asia.202400154

Agboola, O., Fayomi, O. S. I., Ayodeji, A., Ayeni, A. O., Alagbe, E. E., Sanni, S. E., Okoro, E. E., Moropeng, L., Sadiku, R., Kupolati, K. W., & Oni, B. A. (2021). A review on polymer nanocomposites and their effective applications in membranes and adsorbents for water treatment and gas separation. *Membranes, 11*(2), 139. https://doi.org/10.3390/membranes11020139

Aguilar-Pérez, K. M., Ruiz-Pulido, G., Medina, D. I., Parra-Saldivar, R., Nazish, N., Bilal, M., & Iqbal, H. M. N. (2022). Nanoadsorbents for environmental remediation of polluting agents. In *Nano-Biosorbents for Decontamination of Water, Air, and Soil Pollution* (pp. 227–239). Elsevier. https://doi.org/10.1016/B978-0-323-90912-9.00010-1

Alatalo, S.-M., & Sillanpää, M. (2020). Hydrothermal carbonization in the synthesis of sustainable porous carbon materials for water treatment. In *Advanced Water Treatment* (pp. 445–503). Elsevier. https://doi.org/10.1016/B978-0-12-819216-0.00006-0

Alcántara, A. R. (2019). Biocatalysis and pharmaceuticals: A smart tool for sustainable development. *Catalysts, 9*(10), 792. https://doi.org/10.3390/catal9100792

Alcántara, A. R., Domínguez de María, P., Littlechild, J. A., Schürmann, M., Sheldon, R. A., & Wohlgemuth, R. (2022). Biocatalysis as key to sustainable industrial chemistry. *ChemSusChem, 15*(9). https://doi.org/10.1002/cssc.202102709

Alonso, M., Ledesma, B., Román, S., & Olivares-Marín, M. (2024). Insights about the formation of secondary char during HTC processes. *Sustainable Chemistry and Pharmacy, 37*, 101420. https://doi.org/10.1016/j.scp.2023.101420

Babeker, T. M. A., & Chen, Q. (2021). Heavy metal removal from wastewater by adsorption with hydrochar derived from biomass: Current applications and research trends. *Current Pollution Reports, 7*(1), 54–71. https://doi.org/10.1007/s40726-020-00172-2

Baghurst, D. R., & Mingos, D. M. P. (1988). Application of microwave heating techniques for the synthesis of solid state inorganic compounds. *Journal of the Chemical Society, Chemical Communications, 12*, 829. https://doi.org/10.1039/c39880000829

Bhardwaj, N., Agrawal, K., Kumar, B., & Verma, P. (2021). Role of enzymes in deconstruction of waste biomass for sustainable generation of value-added products. In *Bioprospecting of Enzymes in Industry, Healthcare and Sustainable Environment* (pp. 219–250). Springer Singapore. https://doi.org/10.1007/978-981-33-4195-1_11

Cai, T., Du, H., Liu, X., Tie, B., & Zeng, Z. (2021). Insights into the removal of Cd and Pb from aqueous solutions by NaOH–EtOH-modified biochar. *Environmental Technology and Innovation, 24*, 102031. https://doi.org/10.1016/j.eti.2021.102031

Cerón-Camacho, R. (2024). Recent advances in using microwaves to prepare chemicals at the industrial level. *Current Microwave Chemistry, 11*(1), 58–60. https://doi.org/10.2174/01221333562935612402l2114651

Chen, B., Zhou, D., & Zhu, L. (2008). Transitional adsorption and partition of nonpolar and polar aromatic contaminants by biochars of pine needles with different pyrolytic temperatures. *Environmental Science & Technology, 42*(14), 5137–5143. https://doi.org/10.1021/es8002684

Chen, C., Liang, W., Fan, F., & Wang, C. (2021). The effect of temperature on the properties of hydrochars obtained by hydrothermal carbonization of waste *Camellia oleifera* shells. *ACS Omega, 6*(25), 16546–16552. https://doi.org/10.1021/acsomega.1c01787

Chouikhi, N., Cecilia, J. A., Vilarrasa-García, E., Serrano-Cantador, L., Besghaier, S., Chlendi, M., Bagane, M., & Castellón, E. R. (2021). Valorization of agricultural waste as a carbon materials for selective separation and storage of CO2, H2 and N2. *Biomass and Bioenergy, 155*, 106297. https://doi.org/10.1016/j.biombioe.2021.106297

Czerwińska, K., Śliz, M., & Wilk, M. (2022). Hydrothermal carbonization process: Fundamentals, main parameter characteristics and possible applications including an effective method of SARS-CoV-2 mitigation in sewage sludge. A review. *Renewable and Sustainable Energy Reviews, 154*, 111873. https://doi.org/10.1016/j.rser.2021.111873

Devnath, B., Khanal, S., Shah, A., & Reza, T. (2024). Influence of hydrothermal carbonization (HTC) temperature on hydrochar and process liquid for poultry, swine, and dairy manure. *Environments*, *11*(7), 150. https://doi.org/10.3390/environments11070150

Díaz-Rasero, J. M., Ledesma, B., Alonso, M., & Román, S. (2024). Upgrading carthamus by HTC: Improvement of combustion properties. *Fire*, *7*(4), 106. https://doi.org/10.3390/fire7040106

Emenike, P. C., Omole, D. O., Ngene, B. U., & Tenebe, I. T. (2016). Potentiality of agricultural adsorbent for the sequestering of metal ions from wastewater. *Global Journal of Environmental Science and Management*, *2*(4), 411–442. https://doi.org/10.22034/gjesm.2016.02.04.010

Fiori, L., Basso, D., Castello, D., & Baratieri, M. (2014). Hydrothermal carbonization of biomass: Design of a batch reactor and preliminary experimental results. *Chemical Engineering Transactions*, *37*, 55–60. https://doi.org/10.3303/CET1437010

García, A. J. (2011). Surface modification of biomaterials. In *Principles of Regenerative Medicine* (pp. 663–673). Elsevier. https://doi.org/10.1016/B978-0-12-381422-7.10036-7

Goor, O. J. G. M., Keizer, H. M., Bruinen, A. L., Schmitz, M. G. J., Versteegen, R. M., Janssen, H. M., Heeren, R. M. A., & Dankers, P. Y. W. (2017). Efficient functionalization of additives at supramolecular material surfaces. *Advanced Materials*, *29*(5). https://doi.org/10.1002/adma.201604652

Gujjala, L. K. S., Kundu, D., Dutta, D., Kumar, A., Bal, M., Kumar, A., Singh, E., Mishra, R., Kumar, S., & Vo, D.-V. N. (2024). Advances in ionic liquids: Synthesis, environmental remediation and reusability. *Journal of Molecular Liquids*, *396*, 123896. https://doi.org/10.1016/j.molliq.2023.123896

Hassan, M., Liu, Y., Naidu, R., Parikh, S. J., Du, J., Qi, F., & Willett, I. R. (2020). Influences of feedstock sources and pyrolysis temperature on the properties of biochar and functionality as adsorbents: A meta-analysis. *Science of the Total Environment*, *744*, 140714. https://doi.org/10.1016/j.scitotenv.2020.140714

Illankoon, W. A. M. A. N., Milanese, C., Karunarathna, A. K., Liyanage, K. D. H. E., Alahakoon, A. M. Y. W., Rathnasiri, P. G., Collivignarelli, M. C., & Sorlini, S. (2023). Evaluating sustainable options for valorization of rice by-products in Sri Lanka: An approach for a circular business model. *Agronomy*, *13*(3), 803. https://doi.org/10.3390/agronomy13030803

Ippolito, J. A., Cui, L., Kammann, C., Wrage-Mönnig, N., Estavillo, J. M., Fuertes-Mendizabal, T., Cayuela, M. L., Sigua, G., Novak, J., Spokas, K., & Borchard, N. (2020). Feedstock choice, pyrolysis temperature and type influence biochar characteristics: A comprehensive meta-data analysis review. *Biochar*, *2*(4), 421–438. https://doi.org/10.1007/s42773-020-00067-x

Jain, A., Jayaraman, S., Balasubramanian, R., & Srinivasan, M. P. (2014). Hydrothermal pre-treatment for mesoporous carbon synthesis: Enhancement of chemical activation. *Journal of Materials Chemistry A*, *2*(2), 520–528. https://doi.org/10.1039/C3TA12648J

Jain, N., Johnson, T. A., Joshi, H. C., & Prasad, S. (2008). Agricultural and agro-processing wastes as low cost adsorbents for metal removal from wastewater: A review. *Journal of Scientific & Industrial Research*, *67*. https://www.researchgate.net/publication/233746694

Kirk, O., Damhus, T., Borchert, T. V., Fuglsang, C. C., Olsen, H. S., Hansen, T. T., Lund, H., Schiff, H. E., & Nielsen, L. K. (2004). Enzyme applications, industrial. In *Kirk-Othmer Encyclopedia of Chemical Technology*. Wiley. https://doi.org/10.1002/0471238961.0914042114090512.a01.pub2

Kumar, V. (2024). The ionic liquid: An innovative liquid for achieving a sustainable future in chemistry. In *Innovations in Science and Technology: Shaping a Sustainable Future* (Vol. 1, pp. 37–46). B P International. https://doi.org/10.9734/bpi/mono/978-81-973809-6-9/CH4

Lee, J., Yang, X., Cho, S.-H., Kim, J.-K., Lee, S. S., Tsang, D. C. W., Ok, Y. S., & Kwon, E. E. (2017). Pyrolysis process of agricultural waste using CO2 for waste management, energy recovery, and biochar fabrication. *Applied Energy*, *185*, 214–222. https://doi.org/10.1016/j.apenergy.2016.10.092

Mandal, P., Yadav, M. K., Majumder, A., & Ghosal, P. S. (2024). Application of nanomaterials synthesized using agriculture waste for wastewater treatment. In *Green and Sustainable Approaches Using Wastes for the Production of Multifunctional Nanomaterials* (pp. 305–318). Elsevier. https://doi.org/10.1016/B978-0-443-19183-1.00019-2

Nzediegwu, C., Arshad, M., Ulah, A., Naeth, M. A., & Chang, S. X. (2021). Fuel, thermal and surface properties of microwave-pyrolyzed biochars depend on feedstock type and pyrolysis temperature. *Bioresource Technology*, *320*, 124282. https://doi.org/10.1016/j.biortech.2020.124282

Ogunlalu, O., Oyekunle, I. P., Iwuozor, K. O., Aderibigbe, A. D., & Emenike, E. C. (2021). Trends in the mitigation of heavy metal ions from aqueous solutions using unmodified and chemically-modified agricultural waste adsorbents. *Current Research in Green and Sustainable Chemistry*, *4*, 100188. https://doi.org/10.1016/j.crgsc.2021.100188

Osuoha, J. O., & Nwaichi, E. O. (2021). Enzymatic technologies as green and sustainable techniques for remediation of oil-contaminated environment: State of the art. *International Journal of Environmental Science and Technology*, *18*(5), 1299–1322. https://doi.org/10.1007/s13762-020-02876-w

Pandhurnekar, C., Pandhurnekar, H., & Yadao, B. (2024). Microwave-assisted synthesis of biodiesel and related fuel additives. In *The Production of Biodiesel and Related Fuel Additives* (pp. 221–258). Bentham Science Publishers. https://doi.org/10.2174/9789815196740124060010

Polshettiwar, V., Nadagouda, M. N., & Varma, R. S. (2009). Microwave-assisted chemistry: A rapid and sustainable route to synthesis of organics and nanomaterials. *Australian Journal of Chemistry*, *62*(1), 16. https://doi.org/10.1071/CH08404

Rangappa, H. S., Herath, I., Lin, C., & Ch, S. (2024). Industrial waste-based adsorbents as a new trend for removal of water-borne emerging contaminants. *Environmental Pollution*, *343*, 123140. https://doi.org/10.1016/j.envpol.2023.123140

Rangel, M. do C., Mayer, F. M., Carvalho, M. da S., Saboia, G., & de Andrade, A. M. (2023). Selecting catalysts for pyrolysis of lignocellulosic biomass. *Biomass*, *3*(1), 31–63. https://doi.org/10.3390/biomass3010003

Rao, K. J., & Ramesh, P. D. (1995). Use of microwaves for the synthesis and processing of materials. *Bulletin of Materials Science*, *18*(4), 447–465. https://doi.org/10.1007/BF02749773

Roy, S., Bhaskaran, A., & Meenu, P. C. (2023). Microwave-assisted synthesis of porous materials. In *Advances in Microwave-assisted Heterogeneous Catalysis* (pp. 74–91). Royal Society of Chemistry. https://doi.org/10.1039/BK9781837670277-00074

Sarkar, A. K., Kalita, D., & Baruah, S. (2021). Nanomaterials for aquatic contamination sensing and remediation. In *Aquananotechnology* (pp. 67–89). Elsevier. https://doi.org/10.1016/B978-0-12-821141-0.00007-0

Satira, A., Paone, E., Bressi, V., Iannazzo, D., Marra, F., Calabrò, P. S., Mauriello, F., & Espro, C. (2021). Hydrothermal carbonization as sustainable process for the complete upgrading of orange peel waste into value-added chemicals and bio-carbon materials. *Applied Sciences*, *11*(22), 10983. https://doi.org/10.3390/app112210983

Schwenke, A. M., Hoeppener, S., & Schubert, U. S. (2015). Synthesis and modification of carbon nanomaterials utilizing microwave heating. *Advanced Materials*, *27*(28), 4113–4141. https://doi.org/10.1002/adma.201500472

Sethi, S. (2024). Agro waste based adsorbents: Application in cleaning of pharmaceuticals and personal care products from wastewater. In *Development in Wastewater Treatment Research and Processes* (pp. 115–140). Elsevier. https://doi.org/10.1016/B978-0-443-19207-4.00006-9

Shah, S., Dijkstra, J. W., & Wray, H. E. (2024). Process evaluation of mild hydrothermal carbonization to convert wet biomass residue streams into intermediate bioenergy carriers. *Biomass and Bioenergy*, *181*, 107036. https://doi.org/10.1016/j.biombioe.2023.107036

Siddeeg, S. M., Tahoon, M. A., Alsaiari, N. S., Shabbir, M., & Rebah, F. B. (2020). Application of functionalized nanomaterials as effective adsorbents for the removal of heavy metals from wastewater: A review. *Current Analytical Chemistry*, *17*(1), 4–22. https://doi.org/10.2174/1573411016999200719231712

Singh, P., Singh, V. K., Singh, R., Borthakur, A., Madhav, S., Ahamad, A., Kumar, A., Pal, D. B., Tiwary, D., & Mishra, P. K. (2020). Bioremediation. In *Abatement of Environmental Pollutants* (pp. 1–23). Elsevier. https://doi.org/10.1016/B978-0-12-818095-2.00001-1

Singhania, R. R., Patel, A. K., Thomas, L., Goswami, M., Giri, B. S., & Pandey, A. (2015). Industrial enzymes. In *Industrial Biorefineries & White Biotechnology* (pp. 473–497). Elsevier. https://doi.org/10.1016/B978-0-444-63453-5.00015-X

Sun, P., Zhang, K., Shang, S., Song, J., & Wang, D. (2016). Sustainable production of activated carbon spheres from ethyl cellulose. *RSC Advances*, *6*(98), 95656–95662. https://doi.org/10.1039/C6RA16737C

Ullah, Q., Fatima, F., Memon, S. U. R., Khan, M. A., Ismail, A., Rasool, M., Ali, H., Qasim, M., & Ullah, U. (2024). Nanotechnological innovations in agrochemicals: Enhancing efficacy and environmental stewardship in pesticide and herbicide applications. *Middle East Research Journal of Agriculture and Food Science*, *4*(05), 164–178. https://doi.org/10.36348/merjafs.2024.v04i05.002

Valli Sowbhagyam, D. (2024). Ionic liquids as green solvents: A comprehensive review. *International Research Journal on Advanced Engineering Hub (IRJAEH)*, *2*(02), 220–224. https://doi.org/10.47392/IRJAEH.2024.0035

Vermelho, A. B., Supuran, C. T., & Guisan, J. M. (2012). Microbial enzyme: Applications in industry and in bioremediation. *Enzyme Research*, *2012*, 1–2. https://doi.org/10.1155/2012/980681

Wahab, A., Muhammad, M., Ullah, S., Abdi, G., Shah, G. M., Zaman, W., & Ayaz, A. (2024). Agriculture and environmental management through nanotechnology: Eco-friendly nanomaterial synthesis for soil-plant systems, food safety, and sustainability. *Science of the Total Environment*, *926*, 171862. https://doi.org/10.1016/j.scitotenv.2024.171862

Wang, P., Wang, R., & Matulis, V. E. (2024). Ionic liquids as green and efficient desulfurization media aiming at clean fuel. *International Journal of Environmental Research and Public Health*, *21*(7), 914. https://doi.org/10.3390/ijerph21070914

Wang, Z., Qin, X., Dong, H., Liang, Y., Huo, Z., Qian, K., & Yang, F. (2023). Applications of ionic liquids in the field of agriculture: A review. *Agriculture*, *13*(12), 2279. https://doi.org/10.3390/agriculture13122279

Yoshitake, H. (2010). Design of functionalization and structural analysis of organically-modified siliceous oxides with periodic structures for the development of sorbents for hazardous substances. *Journal of Materials Chemistry*, *20*(22), 4537. https://doi.org/10.1039/b927567c

10 Batch and Continuous Biosorption Process and Its Scale-Up to Industrial Application

S Aishwaryalakshmi, S. K. Sivashree, Srinivasan Krishnamoorthy, R Jagan Mohan, N Baskaran, and S Vignesh

10.1 INTRODUCTION

In recent decades, the rapid expansion of industrialization has significantly increased the demand for natural resources, exacerbating the urgent environmental pollution challenges. Inorganic ions, radioactive isotopes, heavy metals (HMs), organometallic compounds, gaseous pollutants, and other organic waste materials are some of the contaminants. The term "heavy metal" refers to metallic elements and metalloids, such as lead, zinc, copper, arsenic, manganese, iron, cobalt, nickel, tin, platinum, gold, mercury, and vanadium, which are hazardous to the environment and ecosystems. Human activity has led to the emergence of HM pollution. Interestingly, metals are not biodegradable and cannot decompose. While certain metals are vital to life and required for certain physiological and biochemical processes, excessive exposure to certain metals can be harmful. Several sources of pollution related to food and agriculture include irrigation with wastewater, pesticides, fertilizers, animal dung, mineral waste, sewage, gasoline spills, coal residues, and trash disposal. Untreated wastewater use has led to a build-up of HMs in our agricultural areas, which are then absorbed by crops that are often consumed by humans. In the meanwhile, accidents involving cars carrying hazardous chemicals and floods that bring sewage and contaminated water to the land cause non-intentional pollution (Briffa et al., 2020).

The HMs are not biodegradable; microbes or chemicals cannot break them down and can remain in the soil for a very long time. The biodegradability of organic pollutants is also impacted by HMs, which reduces their biodegradability and doubles the impact of environmental contamination. Plants taking these metals risk the biosphere and the plant–food chain. Therefore, the scientific community works with legislators to set relevant goals that benefit people and ecosystems and align with the UN's Sustainable Development Goals to address such growing issues in maintaining pollution-free, healthy food ecosystems (Mazarji et al., 2021).

DOI: 10.1201/9781003767251-10

These features of HMs and their possible damage to human health are described in depth. *Lead*: The chemical nature of lead affects its toxicity; approximately 90% of it is absorbed by the body. Organic lead tends to build up mainly in the bones, causing different alterations in the marrow. Children are particularly susceptible to digestive tract absorption of inorganic lead. The respiratory route is thought to present the most significant risk of entry. The central nervous system is affected first by anemia in cases of lead poisoning. In the last stage, there is unconsciousness, kidney failure, seizures, and finally death. *Cadmium*: Clams, crabs, and mushrooms are among the many living things that contain it. Consuming foods high in cadmium, a poor diet, and a lack of vitamin D can lead to osteomalacia (bone softening). The primary sources of contamination from cadmium include pork, fish, potatoes, milk, and beer (Briffa et al., 2020). *Copper* deficiencies cause diseases such as Menkes disease or Wilson's disease. Food poisoning due to copper rarely occurs when the food product is highly acidic, and after a long exposure time, it causes the metal to dissolve. *Zinc* is present in various foods like fish, red meat, grains, dairy products, and cereals. Excessive quantities lead to digestive and metabolic issues. Arsenic exists as arsenite or arsenate, both toxic to humans and animals. It comes from natural sources. Children are most vulnerable, followed by women and men. Typically, neurological issues are more severe in children than in women and, lastly, men. The primary exposure route is through ingesting arsenic-contaminated water. *Chromium*: Found in industrial equipment and fertilizers, its toxicity can lead to burns, anemia, physiological abnormalities, allergic reactions, and malignancies of the gastrointestinal and respiratory systems. Lung cancer may result from breathing in chromate. One of the main ways that people come into contact with chromate is through drinking water (Scutarașu & Trincă, 2023).

10.2 CONVENTIONAL HEAVY METAL REMEDIATION MEASURES

Inorganic wastewater can be treated conventionally with various techniques to remove HMs, such as chemical precipitation, coagulation and flocculation, adsorption, ion exchange, solvent extraction, neutralization, membrane separation, and electrochemical procedures (Yadav et al., 2021).

10.2.1 Physical Methods

10.2.1.1 Membrane Filtration

A driving force is applied across a membrane in membrane filtration, a process for separation that separates substances. Size, solution concentration, applied pressure, and membrane pore size are the factors that this approach uses to differentiate between substrates. The development of membranes driven by high pressure has garnered much attention lately. With an impact on surface charge, hydrophilicity, water permeability, porosity, and thermal/mechanical stability, the material used is critical to the membrane's performance. Removing HMs from wastewater by reverse osmosis is becoming more popular, but it frequently needs high pressure and energy. The removal of HMs from a mixture of mixtures is accomplished by a permeable membrane in ultrafiltration (UF). Two primary approaches have been developed to overcome the drawbacks of metal ion removal by UF membranes: micellar-enhanced ultrafiltration (MEUF) and polymer-enhanced ultrafiltration (PEUF). Separating HMs from wastewater using MEUF is more successful than PEUF (Xiang et al., 2022). Various porous and thin-film composite membranes are used in wastewater treatment, with the help of concepts like osmotic pressure, high pressure, and low pressure. Pore size, distribution, hydrophilicity, surface charge, solution flow, and functional groups are important parameters that influence membrane applications. These components are essential for comprehending the membrane's overall performance in terms of water production rates and the effectiveness of HM removal. This technology has drawn more attention because of its great efficiency and affordability and is now the go-to approach for removing HMs (Teow et al., 2021).

10.2.1.2 Coagulation and Flocculation

Flocculation and coagulation can efficiently remove HMs from sample matrices by lowering the net surface charge of colloidal particles and stabilizing them via electrostatic repulsion. Particles are essentially neutralized in charge during this procedure. Subsequent collisions and interactions with dissolved organic and inorganic polymers gradually combine destabilized particles to make them larger. Filtration, flotation, or straining techniques can effortlessly separate the larger particles that have aggregated from the smaller ones (Fu & Wang, 2011).

10.2.1.3 Ion Exchange

Ion exchange is an economical method that draws soluble ions from the liquid phase into the solid phase using inexpensive materials and an easy-to-use interface. HMs from aqueous solutions have been successfully eliminated in areas with few metal concentrations. The movement of ions from a solution to an immobile solid particle results from this reversible chemical reaction. Industries have successfully implemented the ion exchange approach to eliminate HMs from effluents. But this method is not good for getting rid of many metal ions because of how sensitive the matrix is (Scutarașu & Trincă, 2023).

10.2.1.4 Adsorption

The surface-based adsorption approach accomplishes the extraction of contaminants or adsorbates from wastewater or solutions. By employing this method, impurities stick to the surface of the adsorbent instead of integrating into its crystal structure. Adsorption is a very practical, cost-effective, and ecologically safe method of reducing biological and chemical sludge compared to conventional approaches. Both adsorbent regeneration and metal recovery are feasible. Metal ions in wastewater bind to the surfaces of nanoporous adsorbents during the adsorption process. These adsorbents have enormous surface areas due to their porosity. This process might favorably target one or more metals over others. Regeneration can be achieved with the use of a desorbing agent. Various adsorbents are utilized, including carbon-based adsorbents (high cost), chitosan-based adsorbents (low mechanical strength and stability), and mineral adsorbents (zeolite, silica), where removal efficiency may decline after several cycles, and magnetic adsorbents characterized by magnetic fields, surface charge, and redox activity. Adsorption has been thoroughly studied as a highly promising technique for removing HM ions from wastewater because of its simple operation, wide applicability, high removal rates, and inexpensive reusability. Future research should focus on the most effective approaches for obtaining efficient metal recovery at low prices and with minimal environmental impact. These techniques are already under development (Yadav et al., 2021).

10.2.1.5 Electrocoagulation

Electrocoagulation (EC) is an established technology dating back to the 19th century, but it has not been widely utilized due to substantial capital costs and high electricity demands. Recently, interest in electrochemical technologies has increased because they are environmentally friendly. EC is an electrochemical process involving a reactive anode and cathode, typically made of iron or aluminum. When current flows through the system from a power supply, metallic ions dissolve from the anode and enter the bulk solution. These metallic ions then aggregate into larger flocs, which can be easily removed. The effectiveness of EC can vary considerably, and it produces sludge containing the extracted HMs. However, the energy consumption associated with EC can be quite significant (Tezcan Un & Öcal, 2015).

In an electrochemical system, reduction happens at the cathode (negative side), after electrons are transferred from the anode (positive side). These two chemical reactions are redox (reduction-oxidation), contributing to water purification through metal removal. The type of electrochemical process and the effectiveness of removing particular metal ions are primarily determined by the selection of anode and cathode. The four basic categories of electrochemical processes are electrooxidation, electroporation (ER), EC, and electrolocation (EF). Targeted atoms or molecules are deposited onto the surface of the cathode using the ER technique, also known as electrodeposition or electroplating. The results of the EC method's progressive dissolution of cations from the anodic metal are phase separation, aggregation stability, flotation, precipitation, and the creation of hydroxy complexes. On the other hand, electrode passivation and comparatively high energy requirements are disadvantages of EC. Using insoluble electrodes to perform water electrolysis and adding flotation to improve the treatment process is the mainstay of the EF mechanism (Al Ali et al., 2021).

10.2.2 Chemical Methods

10.2.2.1 Neutralization

The precipitation–neutralization method has been used to remove metals collected in sediments. This process starts with the hydrolysis of lime and progresses to partial dissolution, which raises the pH and creates hydroxyl ions (Yadav et al., 2021).

10.2.2.2 Solvent Extraction

Solvent extraction is the most widely accepted and well-established method utilized in the hydrometallurgical processing of cobalt, nickel, copper, zinc, and other elements. There are currently several stable extractants on the market with improved metal ion selectivity (Mcdonald, 1977).

10.2.2.3 Chemical Precipitation

Chemical precipitation is a widely used industrial technology because it works well, is affordable, allows for process control, and can be applied at different temperatures. The process by which HM ions transform into less soluble compounds that can be removed physically through filtration, flotation, sulfide, or hydroxide depends on the pH of the solution. The amount of contamination to be eliminated will depend on its size, density, and surface charge. Coagulation precipitation, another name for chemical precipitation, is a commonly used technique in industry and is regarded as one of the most reliable. This mechanism helps dissolved metal ions settle by converting them into solid particles. Coagulant reagents can precipitate metal ions by changing the pH, electro-oxidizing potential, or coprecipitation, which is then followed by the removal of sediment. When compared to hydroxide precipitation, the sulfide precipitation process produces fewer dissolved solids and has a greater removal efficiency. This approach has proven successful in treating harmful HM ions; nevertheless, the toxicity of sulfur and its high cost represent major drawbacks. Carbonate precipitation is an efficient and ideal substitute for hydroxide precipitation at lower pH levels. Although it typically produces less sludge volume, it can release CO_2 bubbles and needs more reagent to precipitate effectively (Pohl, 2020).

10.2.2.4 Electrochemical Treatment

Finding a practical and affordable way to remove HM from wastewater has required much work. Removing dangerous soluble metal ions from industrial wastewater can be done most effectively using the electrochemical procedure. The aeration provided by electrolysis gasses improves the process's reaction kinetics. This technique exposed a cathode plate, an insoluble anode, and an aqueous solution containing metal ions to electricity. The idea is to employ an electric current to conjugate electrodes, which precipitate HMs as hydroxides in neutralized or mildly acidic electrolytes. Consequently, selecting the right electrode is essential and improves the process efficiency for impurities.

Traditional chemical and physical methods have been employed for a considerable amount of time to address this environmental issue. These techniques are effective in removing hazardous metals to some or all of their extent. HM ions from aqueous wastes are removed using conventional technologies, such as ion exchange, adsorption, solvent extraction, membrane filtering, chemical precipitation, and lime coagulation. Every procedure has benefits and drawbacks. Many of these techniques are expensive to use in the removal of HMs from aqueous effluents because they are often unsuccessful or unfeasible at concentrations of HMs (10–100 mg/L) above the allowable limit of less than 1 mg/L (Mazarji et al., 2021).

Biological approaches have various advantages over chemical solutions when it comes to treating HM-contaminated environments. One of the most dependable and successful methods for absorbing hazardous metals is the use of bacteria. Numerous studies demonstrate how bacterial species respond differently to HMs through biosorption.

10.3 BIOSORPTION

Increasingly, alternative technologies that are more affordable to use are being made possible by laws governing the release of effluent and growing environmental consciousness. Because of its potential applications, microbial biomass has gained a lot of attention and is gradually becoming a hot topic in the field of metal pollution management. It is a viable option for developing cost-effective and environmentally friendly wastewater treatment processes through biotechnology. Biosorption is an alternate technique that makes use of a variety of naturally occurring biological materials, including bacteria. Biosorption is a technique that uses non-living biological materials, like algae or agricultural leftovers, in conjunction with physical and chemical interactions to remove contaminants and HMs from wastewater. Treating industrial wastewater is a cost-effective and ecologically friendly method that is independent of the biomass's metabolic activity.

The ability of biological materials to extract HMs from wastewater by physicochemical or metabolic routes is known as biosorption. The term "bioprocess" also describes how some microbial biomaterials may bind and precipitate HMs from even the tiniest water solutions, offering a workable and financially appealing substitute. The ability of specific biomolecules (or biomass types) to bind and concentrate particular ions or other molecules from aqueous solutions is known as biosorption. Yeasts, fungi, bacteria, and algae have all shown promise as metal biosorbents. Because of their outstanding performance, high availability, and relatively low processing costs, algae are the most widely used biosorbent. Additionally, brown algae offer other benefits that make it a perfect substitute for other methods when it comes to eliminating pollutants, mainly from effluents. Because conventional procedures are costly and environmentally friendly, low-cost materials are needed.

However, biosorption continues to be the most economical and environmentally benign way to eliminate HMs from wastewater used in homes and businesses. It offers numerous benefits, including low cost, excellent efficiency, decreased biological and chemical sludge, and the capacity to recover metals through the regeneration of biosorbents. The biosorption process has a number of desirable characteristics, including inexpensive capital and operating costs, quick kinetics of adsorption and desorption, and the capacity to remove metals over a broad pH and temperature range. Biomass may be readily renewed and repurposed. Most notably, biological biomass has a higher capability for removing metals than traditional adsorbents (Sulaymon, 2014).

10.3.1 Successful Biosorption Applications in Heavy Metal Removal

One study examined how the blue-green algae *Anabaena sphaerica* removed lead (II) and cadmium (II) from an aqueous solution). According to the calculations, this algae can remove lead (II) and cadmium (II) with comparatively high biosorption capacities of 111.1 and 121.95 mg/g, respectively. Another study examined the capacity of chemically altered *Cystoseira indica* to biosynthesize lead (II) and uranium (VI) oxide. It was discovered that sargassum biomass may biosorb lead and cadmium cations by partially or fully esterifying the carboxylic sites on its cell wall. Currently, research is primarily focused on developing and optimizing advanced biosorbents. This involves using various biological resources, including fungi, algae, microbes, and agricultural waste. Scientists are working hard to find and develop biosorbents with better selectivity and affinity for HMs. This approach seeks to lessen ecological impact while increasing the effectiveness of pollution removal from aqueous solutions. It has been proposed that coexisting pollutants, such as combined cobalt ions and Congo red dye, can be simultaneously removed from model solutions using the naturally dried biomass of the macro-green alga *Enteromorpha intestinalis* (Redha, 2020). *Limosilactobacillus fermentum* CN-005 and *Lactobacillus fermentum* CN-011, two recently discovered lactic acid bacteria strains, have shown a high sorption capacity and Pb (II) tolerance. These strains were utilized to successfully biosorb lead (II) from wastewater that was modeled. Furthermore, demonstrating their ecological and financial significance for the sustainability of ecosystems are fungi well-known for their capacity to absorb and accumulate HMs. Over seven days, *Penicillium* sp. and *A. niger* sorbed 96.4%. However, using fungi as biosorbents is a hard undertaking due to difficulties in the biosorption kinetics and competition with other ions in the environment. Biosorption conditions are typically optimized by a variety of statistical and mathematical techniques, such as central composite design (CCD) in response surface methodology (RSM) or statistical design of experiments (DOE). Combining these methods can help determine ideal circumstances by offering a comprehensive understanding of the relationships between various factors and how they affect biosorption effectiveness. Furthermore, cutting-edge computational methods like genetic algorithms and artificial neural networks successfully improve biosorption process optimization. The use of machine learning methods for biosorption system predictive modeling is another area of research interest (Kanamarlapudi et al., 2018).

10.4 TECHNIQUES FOR ISOLATION OF MORE POTENT HEAVY METAL-RESISTANT AND BIOSORBENT MICROBES

The isolation of HM-resistant strains is essential for developing mechanisms for removing HMs that have detrimental effects on the environment and human health. The following methods can be used to isolate these metal-resistant microbes.

10.4.1 Enrichment and Isolation

In this method, the samples are collected and taken in sterile plastic bottles and transported to the laboratory in an ice box for bacteriological purposes. Then, a nutrient agar medium containing various concentrations of different HM compounds is prepared. A sterile wire loop is used to collect loops full of pure isolates and directly streak on the surface of HMs. The loops are then incubated at 37°C for 24 h. The bacteria can then be isolated and analyzed by determining their optical density over time.

10.4.2 Screening of Biosorbent Microbes

The HM tolerance of the bacterial isolates is determined by increasing the concentrations of HMs until the bacteria fail to give colonies on the plate. A minimum inhibition concentration (MIC) is noted when isolates fail to grow after incubation. The MIC of a HM is the lowest concentration of the metal that stops the growth of a microorganism. Bacterial colonies were streaked onto agar plates containing increasing HM concentrations. The plates were incubated at appropriate temperatures for bacteria, and the lowest concentration of the HM that prevents the organism from growing is considered as MIC of the HM. MICs can be determined using agar or broth media, but MICs in solid media are usually higher than those in liquid media (Bender et al., 2022).

10.4.3 Molecular Identification of Metal-Tolerant Biosorbent Strains

The bacterial genomic DNA of HM-tolerant strains are isolated and amplified by using the universal bacterial primers, F (5′-CAG GCC TAA CAC ATG CAA GTC-3′) and R (5′-GGG CGG WGT GTA CAA GGC-3′). A typical polymerase chain reaction (PCR) performed with 20 μl reaction mixture containing 1× PCR buffer (100 mM Tris HCl, pH 8.3; 500 mM KCl), 0.2 mM each dNTPs (dATP, dGTP, dCTP, and dTTP), 2.5 mM $MgCl_2$, 1 unit of AmpliTaq Gold DNA polymerase enzyme, 0.1 mg/ml BSA, 4% DMSO, 5 pM of forward and reverse primers, and template DNA. PCR was performed with an initial denaturation for 5 min at 95 °C for 30 s, annealing at 60 °C for 1 min, extension at 72 °C for 1 min, and a final extension at 72 °C for 7 min. PCR products were analyzed by around 1% (w/v) agarose gel electrophoresis in 0.5× TBE buffer with Ethidium bromide (0.5 μg/ml) (Kanamarlapudi et al., 2018).

10.5 TYPES OF BIOSORBENTS

Different biomaterials with high adsorption, ion exchange, and coagulation properties are considered biosorbents. They are employed in multiple applications to extract impurities from solutions. Physical, chemical, and biological modifications can greatly increase their effectiveness.

10.5.1 Chemically Modified Biosorbents

To increase the biosorption capacity, a variety of pretreatments are applied, such as acetone, ethanol, acid, and alkaline treatments. For example, using 0.1 M HNO_3 to increase metal uptake shows that acidic pretreatments can eliminate contaminants that impede binding sites. Functional compounds such as carboxyl, amine, or hydroxyl groups can also be added chemically to boost binding sites. Metal absorption capabilities can be significantly increased, for instance, by adding carboxyl groups via chloroacetic acid or by altering biomass with citric acid (Bender et al., 2022).

10.5.2 Genetically Modified Biosorbents

Genetic manipulation results in biosorbents with enhanced selectivity and capacity. Microorganisms' ability to sequester metal can be greatly enhanced by designing them to overexpress peptides such as phytochelatins or metallothioneins. For instance, recombinant strains of Staphylococcus species or *E. coli* have demonstrated increased metal uptake by expressing metal-binding proteins or peptides on their surfaces.

10.5.3 Immobilized Biosorbents

Biosorbents can be immobilized within matrices like sodium alginate, polysulfone, polyacrylamide, or polyurethane to avoid problems like low mechanical strength and challenging biomass separation. Immobilization lowers process costs, promotes reuse, and strengthens the biomass mechanically. Polysulfone matrices, for example, have been useful in preserving high biosorption capabilities and permitting several cycles of use (Adewuyi, 2020).

10.6 BATCH MODE OF BIOSORPTION

One of the most researched and used techniques for eliminating impurities from aqueous solutions is batch biosorption (Volesky & Holan, 1995; Aksu et al., 2002). It involves subjecting a certain quantity of biomass, or biosorbent, to a pollutant-containing solution in a closed system for a predefined amount of time (Table 10.1). During this time, the pollutants are removed from the solution because of interactions between the biomass and them. The procedure continues until the target contaminant

TABLE 10.1
Advantages and Limitations of Batch-Type Biosorption

Advantages	Limitations
1. **High Control Over Experimental Variables**: Batch processes allow precise control over factors such as pH, temperature, and contact time, enabling the optimization of biosorption conditions (Aksu, 2005)	1. **Time-Consuming**: Batch processes may require longer time periods to reach equilibrium, especially when dealing with low concentrations of contaminants (Aksu et al., 2002)
2. **Cost-Effective**: Batch biosorption is relatively inexpensive as it uses low-cost biosorbents like microbial biomass, agricultural waste, or industrial by-products	2. **Limited Scalability**: While batch processes work well in laboratory settings, scaling up to industrial levels can be challenging and inefficient
3. **High Efficiency for Low Pollutant Concentrations**: This method is especially effective for treating low concentrations of contaminants, which can be challenging for conventional methods	3. **Difficult Regeneration**: Biosorbent regeneration can be challenging, especially if the contaminants are strongly bound, making the process less efficient for reuse
4. **Simple Operation**: The process is straightforward, requiring basic laboratory equipment, such as beakers, stirrers, and filtration units (Singha et al., 2020)	4. **Limited Capacity**: The biosorbent in a batch system can quickly become saturated, requiring frequent replacement or regeneration (Vijayaraghavan & Yun, 2008)
5. **Wide Range of Biosorbents**: A variety of biosorbents can be used, including microbial biomass, algae, fungi, and plant materials (Aksu, 2005)	5. **Inconsistent Results**: Results may vary depending on contaminant concentration, biosorbent preparation, and process conditions, making reproducibility a challenge (Aksu et al., 2002)
6. **Effective for Multiple Pollutants**: Batch biosorption can simultaneously remove multiple pollutants from the solution, such as heavy metals and dyes (Volesky & Holan, 1995)	6. **Waste Generation**: Spent biosorbents, especially those not easily regenerated, can lead to secondary waste disposal issues (Volesky & Holan, 1995)

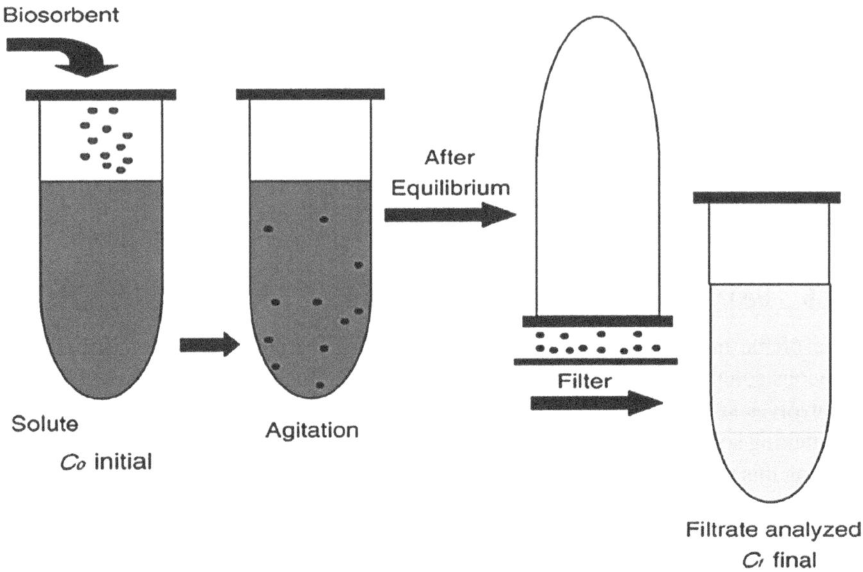

FIGURE 10.1 Batch-type biosorption process.

concentration level is reached, equilibrium is reached, or no further adsorption takes place. For example, the impact of temperature and pH on metal uptake was demonstrated in the batch biosorption experiment with wheat straw biomass to adsorb copper and cadmium ions from aqueous solutions. Control variables like mixing speed, temperature, and initial metal ion concentrations helped determine optimal biosorption conditions (Qasem et al., 2021). The metal uptake was calculated using standard formulas, and the experimental error was within acceptable limits, demonstrating the technique's efficiency for industrial applications. This method involves mixing pollutants and biosorbents in a single vessel. The system keeps running until it reaches equilibrium, at which point the biosorbent absorbs contaminants in the solution. This method is simple and appropriate for small-scale applications (Figure 10.1).

10.6.1 Formula for Metal Uptake

The following formula was used to determine the quantity of metal absorbed by the biosorbent (q, in mg/g) (Ho & Mckay, 1999) (Aksu et al., 2002):

$$q = \frac{(c_0 - c_t)V}{M}$$

where:

q: metal uptake (mg g^{-1}),
c_0: initial concentration (mg L^{-1}),

c_t: metal concentration at a given time (mg L^{-1}),
M: mass of biosorbent (g), and
V: volume of the solution (L).

10.6.2 Steps Involved in Batch Mode Biosorption

- **Preparation of Biomass**: Biosorbents can be made from a variety of biomass types, including dried, chemically processed, or living or dead cell biomass. To increase the material's capacity for biosorption, the preparation may involve pretreatment, drying, or grinding (Volesky & Holan, 1995; Aksu et al., 2002).
- **Preparation of the Contaminant Solution**: The contaminant (such as HMs or dyes) is produced in a solution with a defined concentration. To guarantee uniformity in the results, variables including pH, temperature, and the concentration of the original contamination are controlled (Ho & Mckay, 1999).
- **Biosorption Process**: The processed biomass is added to the contaminated solution within a closed vessel, usually a flask or beaker. After that, the mixture is agitated or mixed to ensure that the contaminant and biomass are properly in touch. Higher removal rates and improved interaction are encouraged by this (Vijayaraghavan & Yun, 2008).
- **Time Duration**: The length of time the process is let on is determined by the biosorption kinetics and usually ranges from a few minutes to many hours. Samples are taken periodically over this period to track the contaminant's concentration in the solution (Aksu, 2005).
- **Separation and Analysis**: Once the required biosorption time has passed, the biosorbent is extracted from the solution, usually by filtration or centrifugation. The residual concentration of the pollutant in the solution is examined to ascertain the biosorption effectiveness (Volesky & Holan, 1995).

10.6.3 Factors Affecting Batch Biosorption

The efficiency of biosorption is influenced by several key factors (Figure 10.2):

- **Solution pH**: The chemistry of the solute (such as metal speciation) and the activity of the functional groups in the biomass are two factors that pH influences, which are critical to biosorption. For example, the ion exchange process with carboxyl groups makes pH 3–6 ideal for metal ion biosorption. Based on their class, they require varying pH values.
- **Temperature**: Temperature-induced increases in surface activity and kinetic energy lead to a general improvement in biosorption efficiency. High temperatures, however, can potentially harm the biosorbent and lower its capacity for biosorption in exothermic processes.
- **Ionic Strength**: Stronger ions can compete with one another and alter the electrical double layer around the biomass's surface, reducing biosorption.

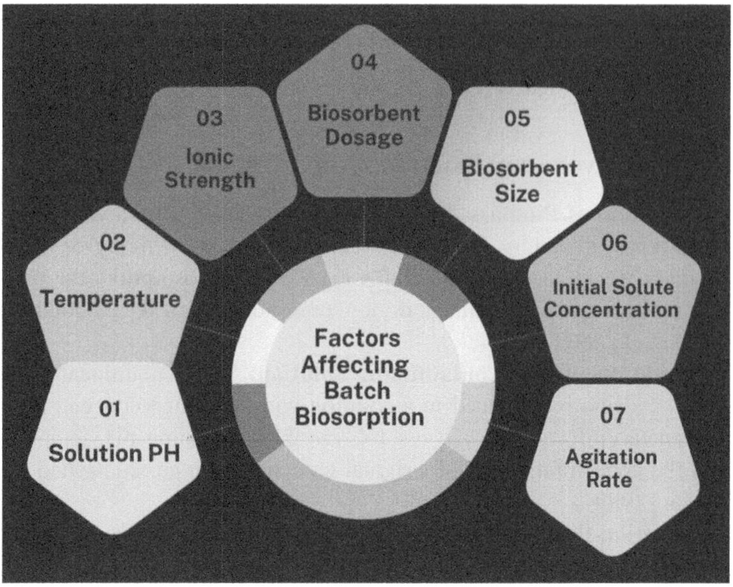

FIGURE 10.2 Factors affecting batch-type biosorption process.

- **Biosorbent Dosage**: While higher concentrations of the biosorbent usually result in greater overall absorption, site saturation can cause poorer specific absorption at high dosages.
- **Biosorbent Size**: Smaller particles provide a higher surface area, allowing for faster biosorption. Mechanical robustness is crucial for practical applications.
- **Initial Solute Concentration**: As binding sites get saturated, higher solute concentrations promote biosorption but also decrease removal efficiency (Ho & Mckay, 1999).
- **Agitation Rate**: Increasing agitation enhances mass transfer by decreasing the thickness of the boundary layer surrounding biosorbent particles and speeding up solute diffusion.

10.7 BIOSORPTION PROCESS IN A CONTINUOUS MODE

In biosorption, a continuous mode of operation involves passing a contaminated solution through a fixed bed or packed column containing biosorbent material to remove pollutants, typically HMs or dyes. This mode mimics real-world industrial wastewater treatment systems, making it more suitable for large-scale applications than batch systems. Continuous biosorption offers the benefit of maintaining constant flow conditions, reducing downtime, and enhancing the system's scalability (Figure 10.3). The standard configuration for continuous biosorption is a packed column, in which the biosorbent is packed into a cylindrical column and the polluted solution is forced through at a regulated flow rate. Depending on the intended interaction between the

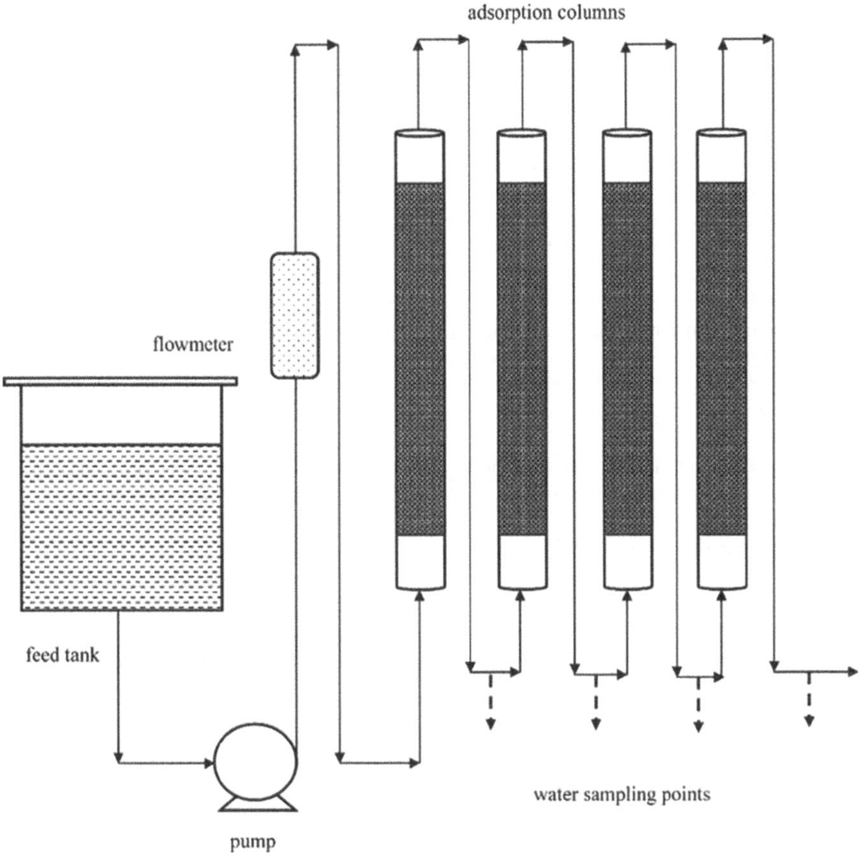

FIGURE 10.3 Continuous-type biosorption process.

biosorbent and the pollutant, the column can function in either upward or downward flow modes (Table 10.2). For example, a continuous flow biosorption experiment using wheat straw to remove copper and cadmium ions involved pumping contaminated water through columns packed with biosorbent material at different bed heights and flow rates. Breakthrough curve analysis was conducted to evaluate adsorption kinetics, helping to optimize the system for effective pollutant removal. This scalable technique is ideal for industrial applications requiring continuous treatment of wastewater (Volesky & Holan, 1995; Aksu, 2005; Vijayaraghavan & Yun, 2008).

The process of continuous biosorption is delineated in the following steps:

- **Column Design**: The biosorbent, which can be bacterial biomass, agricultural waste, or other biological materials, is typically put into a cylindrical column. The pollutant concentration and the intended removal efficiency determine the column's height, diameter, and biosorbent content (Volesky & Holan, 1995; Vijayaraghavan & Yun, 2008).

TABLE 10.2
Advantages and Limitations of Continuous-Type Biosorption

Advantages	Limitations
1. **Scalable for Large-Scale Operations**: Continuous systems are more suitable for industrial wastewater treatment, as they allow for large volumes of water to be treated in a shorter time than batch processes (Volesky & Holan, 1995; Vijayaraghavan & Yun, 2008)	1. **Column Saturation**: The biosorbent becomes saturated over time, gradually decreasing removal efficiency. Frequent regeneration or replacement is needed (Volesky & Holan, 1995; Aksu et al., 2002)
2. **Efficient Use of Biosorbent**: The biosorbent is fully utilized in a continuous mode as the solution continuously flows through the column, maximizing adsorption efficiency (Aksu, 2005)	2. **Breakthrough Time**: Breakthrough occurs when the contaminant begins to appear in the effluent, requiring careful monitoring to prevent reduced performance
3. **Economical and Time-Saving**: Continuous operation reduces downtime associated with batch processes, such as emptying and refilling the reactor (Aksu, 2005)	3. **Complex Column Design**: Designing the column for optimal flow rates, pressure drops, and biosorbent distribution can be complex and costly (Aksu et al., 2002)
4. **Real-Time Monitoring**: The process can be monitored in real-time, allowing adjustments to flow rates, pH, and other parameters to optimize contaminant removal (Volesky & Holan, 1995)	4. **Regeneration Challenges**: Regenerating the biosorbent can be time-consuming and may require harsh chemicals, which could lead to biosorbent degradation (Volesky & Holan, 1995)
5. **Consistent Effluent Quality**: Continuous operation provides a more consistent effluent quality as the process runs without interruption (Vijayaraghavan & Yun, 2008)	5. **Risk of Clogging**: Continuous systems, particularly with particulate biosorbents, may face clogging issues if the feed solution contains high levels of suspended solids

- **Feed Solution**: A solution containing the pollutant is continually added at the top or bottom of the column. To guarantee that the biosorbent and the pollutant make uniform contact, the solution flows at a steady flow rate (Aksu et al., 2002).
- **Biosorption Process**: The contaminant attaches itself to the biosorbent material by a variety of mechanisms, including ion exchange, complexation, and surface adsorption, as the solution moves along the biosorbent-packed column. Contaminants are still being absorbed by the biosorbent till its limit is reached (Ho & Mckay, 1999).
- **Breakthrough Curve**: A breakthrough curve is created to investigate the adsorption dynamics in continuous biosorption systems. It shows the pollutant content in the wastewater over time. The biosorbent is getting saturated when the contaminant starts to show up in the effluent, which is the breakthrough point.
- **Biosorbent Replacement or Regeneration**: After the biosorbent is saturated, it can either be replaced with new material or regenerated using chemical solutions that desorb the impurities. In order to lower operating costs and increase process sustainability, regeneration is necessary.

10.7.1 Factors Affecting Continuous Biosorption Efficiency (Figure 10.4)

- **Flow Rate**: The amount of time that the biosorbent and the contaminant are in direct contact depends on how quickly the contaminated solution passes through the column. Increased contact time from a slower flow rate could lead to better adsorption efficiency. On the other hand, abnormally low flow rates may result in inefficient operations (Volesky & Holan, 1995).
- **Column Height and Diameter**: The column's dimensions serve a crucial role in establishing how long the solution stays in the system. Longer contact times in taller columns improve adsorption, but they can also cause a pressure reduction across the column (Aksu, 2005).
- **Biosorbent Characteristics**: The biosorbent's adsorption capability is determined by its characteristics, including its surface area, porosity, and functional groups. In continuous systems, certain biosorbents could benefit from pretreatment to increase their adsorption efficiency (Vijayaraghavan & Yun, 2008).
- **pH and Temperature**: The adsorption process can be influenced by the pH of the solution, which can also change the ionization state of the biosorbent

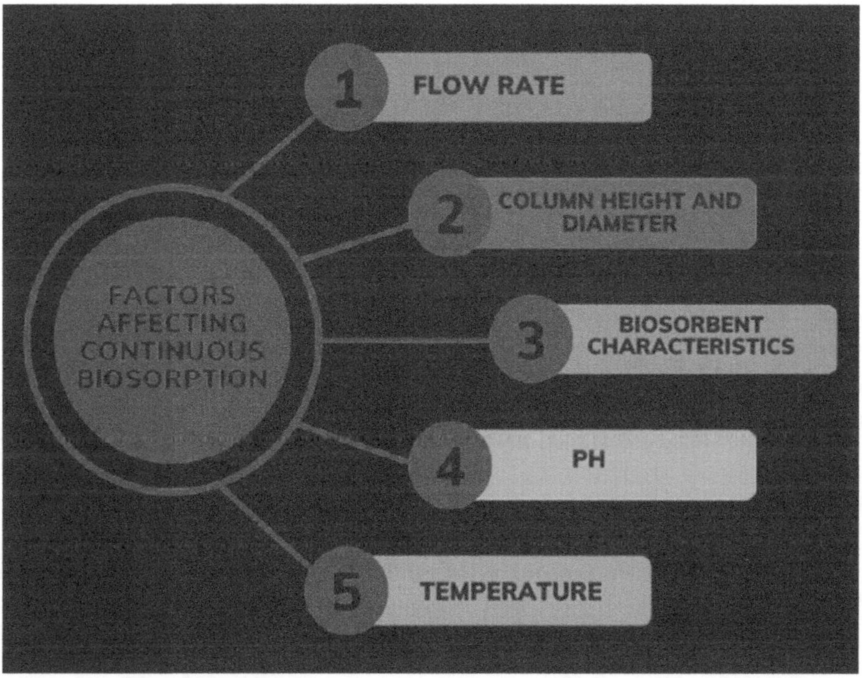

FIGURE 10.4 Factors affecting the continuous-type biosorption process.

and the contaminant. The kinetics of biosorption can also be affected by temperature, with higher temperatures often leading to a faster rate of adsorption (Volesky & Holan, 1995).

10.8 STATISTICAL DESIGNS AND MODELING TOOLS FOR THE BIOSORPTION PROCESS

Systematic experimental methods are necessary to evaluate the performance of various biosorbents. The sorption, kinetics, and equilibrium models are the main topics of the statistical designs of experiments (DOE) and screening methods in biosorption research. DOE improves the understanding and optimization of the biosorption process by reducing the required trials while preserving high accuracy. Selecting a sorption model is essential to explaining how solutes interact with biosorbents. Biosorption investigations have used various models, from simple empirical models to more complex mechanical models. These models are helpful in estimating process outcomes, assessing biosorption capabilities, and determining affinities (Volesky & Holan, 1995; Vijayaraghavan & Yun, 2008).

10.8.1 Langmuir Isotherm

According to the Langmuir isotherm, biosorption occurs at certain homogenous locations within the biosorbent and is assumed to be monolayer adsorption. The following equation shows it:

$$Q = \frac{Q_{\max} b_{\mathrm{L}} C_{\mathrm{f}}}{1 + b_{\mathrm{L}} C_{\mathrm{f}}}$$

Q: solute uptake (mg/g),
$Q_{\max}$: maximum sorption capacity (mg/g),
b_{L}: affinity constant, and
C_{f}: equilibrium solute concentration.

The Langmuir model's constants are helpful when developing adsorption systems based on maximum capacity. It is particularly beneficial for systems where saturation is observed (Volesky & Holan, 1995; Ho & Mckay, 1999).

10.8.2 Freundlich Isotherm

When applied to heterogeneous systems, the Freundlich isotherm assumes that biosorption occurs on surfaces with different affinities. The model can be stated as follows:

$$Q = K_{\mathrm{F}} C_{\mathrm{f}}^{\frac{1}{nF}}$$

K_F: Freundlich constant related to adsorption capacity and
C_f: heterogeneity factor, indicating surface uniformity

This model is frequently used when multi-layer adsorption results from uneven energy distribution at the adsorption sites (Aksu, 2005).

10.8.3 Temkin Isotherm

The impacts of indirect interactions between molecules of the solute and the biosorbent are considered by the Temkin isotherm.

The following is the Temkin model:

$$Q = \frac{RT}{b_T} \ln\left(a_T C_f\right)$$

R: universal gas constant,
T: temperature (Kelvin),
b_T: related to the heat of sorption, and
a_T: Temkin constant

This model is especially helpful when sorption heat varies linearly with coverage due to sorbate–sorbent interactions (Ho & Mckay, 1999).

10.8.4 Redlich–Peterson Isotherm

The Redlich–Peterson model is more versatile since it incorporates Freundlich and Langmuir isotherms elements. The model can be written as follows:

$$Q = \frac{K_{RP} C_f}{1 + a_{RP} C_f^{\beta_{RP}}}$$

K_{RP}, a_{RP}, and β_{RP} are the isotherms constants.

Under various experimental settings, this model can reduce to either the Freundlich model or the Langmuir model depending on the concentration of sorbate (Ho & Mckay, 1999).

10.8.5 Sips Isotherm

The Sips isotherm manages both low- and high-concentration sorption scenarios as a mix of the Freundlich and Langmuir models:

$$Q = \frac{K_S C_f^{\beta_S}}{1 + a_S C_f^{\beta_S}}$$

a_S and β_S are Sips constants.

Since the adsorption capacity rises with concentration but levels off as it approaches saturation, the Sips isotherm is very helpful in biosorption investigations involving heterogeneous systems.

10.8.6 Kinetics in Biosorption Studies

Studying biosorption kinetics is essential to developing effective remediation systems based on biosorbents. Understanding the sorption mechanism, control phases, and potential rate-limiting stages is possible with the help of kinetic models.

10.8.6.1 Pseudo-First-Order Kinetics

The basic idea of the pseudo-first-order model is that the number of vacant biosorption sites equals the rate of site occupation. It is said as follows:

$$q_t = q_e\left(1 - e^{-k_1 t}\right)$$

q_t: Amount of solute adsorbed at time t,
q_e: Equilibrium sorption capacity, and
k_1: Pseudo-first-order rate constant.

This model works especially well in the early stages of sorption, when the available biosorbent sites are rapidly interacting with the sorbate molecules.

10.8.6.2 Pseudo-Second-Order Kinetics

According to the pseudo-second-order model, biosorption happens by chemisorption, which involves valency forces through the sharing or exchange of electrons. It appears as follows:

$$q_t = \frac{k_2 q_e^2 t}{1 + k_2 q_e t}$$

k_2: pseudo-second-order rate constant and
q_e: equilibrium sorption capacity

This model is often a better fit for biosorption data than the pseudo-first-order model, as it accounts for the entire sorption process, not just the initial stages.

10.8.6.3 Intraparticle Diffusion Model

Sorption onto immobilized biosorbents may include numerous phases, with intraparticle diffusion being the rate-limiting step in some circumstances. The model is represented as follows:

$$q_t = k_i t^{\frac{1}{2}}$$

k_i: intraparticle diffusion rate constant.

A plot of q_t against $t^{\frac{1}{2}}$ helps to identify whether intraparticle diffusion is the limiting factor, which is crucial in large-scale industrial applications.

10.8.6.4 Equilibrium Models in Biosorption

Equilibrium models are used to describe how biosorption progresses over time until the system reaches equilibrium. These models are essential for designing and optimizing biosorption systems for practical applications.

10.8.6.5 Langmuir Equilibrium Model

According to the Langmuir model, no lateral interaction exists, and sorbate molecules form a monolayer on a uniform adsorption surface. This model is frequently used when no more sorption occurs when the biosorbent reaches a saturation threshold.

10.8.6.6 Freundlich Equilibrium Model

The Freundlich model, being empirical, is more suited for heterogeneous biosorbents with multiple layers of adsorption. It predicts that as the concentration of the solute increases, biosorption becomes less efficient, reflecting the occupation of lower-energy sorption sites.

10.8.6.7 Temkin and Redlich–Peterson Models

The Temkin and Redlich–Peterson models offer options for challenging biosorption systems where interactions between sorbate molecules influence the sorption process. When other isotherms fall short, such as when there are irregular biosorbents and non-ideal adsorption behavior, these models can provide a better fit to the data.

10.9 STATISTICAL DESIGN OF EXPERIMENTS (DOE) AS A SCREENING TOOL

Optimizing biosorption processes requires careful consideration of experimental design (Table 10.3). Full factorial, fractional factorial, and response surface methods are statistical techniques that facilitate the systematic screening of factors such as pH, temperature, dosage of the biosorbent, and beginning solute concentration (Volesky & Holan, 1995; Aksu, 2005).

10.10 RESPONSE SURFACE METHODOLOGY (RSM)

RSM is often utilized to investigate the relationships between various experimental variables to optimize biosorption processes. It makes it possible to determine ideal operating conditions with the least amount of experimental testing. These studies frequently use Box–Behnken and CCDs.

TABLE 10.3
Factors That Influence the Biosorption Process

Parameter	Range/Levels	Significance
pH	3–9	Controls sorption efficiency
Temperature (°C)	20–60	Affects reaction rate
Biosorbent dosage (g)	0.1–5	Determines capacity
Initial solute concentration (mg/L)	10–1000	Higher concentration, higher sorption
Contact time (h)	1–24	Determines equilibrium time

10.11 BIOSORPTION AS AN ALTERNATIVE AND SUSTAINABLE APPROACH

Biosorption offers diverse applications across several sectors due to its adaptability and effectiveness:

- **Wastewater Treatment**: Biosorption removes dangerous HMs from industrial effluents, including cadmium, lead, arsenic, mercury, and chromium, to reduce water pollution and safeguard aquatic ecosystems.
- **Environmental Remediation**: It helps restore contaminated places, reduces environmental risks, and cleans contaminated soil and groundwater by removing metal contaminants.
- **Recovery of Precious Metals**: Biosorption is a technique used in the mining industry to recover valuable metals from waste streams, including gold, silver, and platinum. This procedure reduces the negative effects of mining operations on the environment while improving resource recovery.
- **Textile and Dye Industries**: Biosorption efficiently addresses the problem of synthetic dye pollution in wastewater from the textile and dye industries, enhancing the quality of the treated water and making it easier to reuse.
- **Food Industries**: Food industries dispose of large quantities of waste and byproducts. Agricultural wastes, especially those with high percentages of cellulose and lignin, contain polar functional groups like amino, carbonyl, alcoholic, phenolic, and ether groups with a high potential for metal binding. Microorganisms capable of tolerating unfavorable conditions evolved their use as biosorbents in the removal of metal ions from wastewater. Table 10.4 lists various biosorbents and their sources.

10.12 FUTURE RESEARCH POTENTIAL IN BIOSORPTION AND ITS ROLE IN PROGRESSING INDUSTRIAL POLLUTION REMEDIATION

Bacterial biomass is one extremely promising class of biosorbents for removing metal ions and dyes from industrial effluents. However, the lack of selectivity in practical applications and restrictions on reprocessing microbial biomass has prevented its

TABLE 10.4
Types of Biosorbents and Their Sources

Type of Biosorbent	Source	Biosorbate
Tea waste	Tea processing industry	Chromium
Fly ash	Cement industry	Lead
Rice husk	Paddy processing industry	Nickel
Sugarcane bagasse	Sugarcane processing industry	Nickel
Date pit	Date processing industry	Copper, cadmium
Mucor rouxii	Algae	Copper
Bacillus cereus	Bacteria	Zinc
Aspergillus niger	Fungi	Copper

widespread use. Despite more than ten years of essential research, only slow progress has been made in the commercialization of biosorption for wastewater treatment. For biosorption technology to reach its full potential, more research is required in several major areas:

- **Biomass Selection and Screening**: It is critical to find and choose the biomass with the highest biosorption capability and selectivity. Extensive research into a variety of species is required to determine which microbial species function best under diverse circumstances (Vijayaraghavan & Yun, 2008).
- **Optimization of Operational Conditions**: To optimize biosorption efficiency, it is critical to adjust factors including pH, temperature, ionic strength, and the presence of co-ions. By comprehending how these variables affect biosorption, more reliable and scalable procedures can be developed (Aksu, 2005).
- **Improving Selectivity and Capacity**: Biomass can be chemically and genetically modified to greatly increase its capacity and ability to absorb pollutants. To maximize metal ion binding, scientists are already investigating techniques, including chemical pretreatments and genetic modification of microbial cells (Wang & Chen, 2009).
- **Enhancing Mechanical Strength and Reusability**: Its mechanical strength must be increased to make biomass useful. Techniques for immobilization or other chemical treatments can increase the biomass's stiffness, preserving its effectiveness over time (Volesky & Holan, 1995).
- **Testing in Real-World Environments**: Extensive testing is necessary to determine how well biosorbents operate in treating industrial effluent. This involves determining how biosorption efficiency is affected by effluent composition and water quality. Although biosorption is a difficult technology to replace existing wastewater treatment methods, the advantages could be significant. Biosorption may address various contaminants and is economical and environmentally friendly. Shortly, biosorption technology

is expected to become a more viable and appealing alternative for treating industrial pollution, provided further research is directed toward pilot and full-scale applications (Vijayaraghavan & Yun, 2008).

10.13 EXAMPLES OF SUCCESSFUL INDUSTRIAL BIOSORPTION APPLICATIONS AND PATENTS

- Biosorbent of HM removal comprising watermelon rind, sugarcane bagasse, and garden grass (WO2014012134A1)
- Biosorbent of HMs using plant waste powder, algae powder, sodium alginate, gelatin, colloidal fluid, organic solvent, and adhesive (CN106563417B)
- Biosorbents for removal of metals from aqueous solutions (AU2011201389B2)

10.14 CONCLUSION

The chapter comprehensively reviewed the batch and continuous biosorption processes for removing HM pollutants from different ecosystems. The mechanism behind the biosorption process was technically explained in the context of adsorption, ion exchange, and other metabolic pathways. Different operational modes of batch and continuous biosorption processes were analyzed with their merits and demerits. In perceiving the importance of the biosorption process through various mediums, scholarly works were collectively presented with different sorption capacity and their influencing variables. Several kinetics and models were described in detail to understand the dynamics of the biosorbents and their mechanisms. The chapter also identified the prevailing research gaps in biosorption and its scalability. From the vast literature reports, this work concludes with successful biosorption applications in the potential HM removal process and future research potentials viz (a) biomass selection and screening, (b) development and optimization of operational conditions for targeting different pollutants, (c) improving selectivity and capacities, and (d) enhancing the stability for more applications. Hence, addressing these research areas scientifically would benefit the preservation of multiple ecosystems.

REFERENCES

Adewuyi, A. (2020). Chemically modified biosorbents and their role in the removal of emerging pharmaceutical waste in the water system. *Water (Switzerland)*, *12*(6). https://doi.org/10.3390/W12061551

Aksu, Z. (2005). Application of biosorption for the removal of organic pollutants: A review. *Process Biochemistry*, *40*(3–4), 997–1026. https://doi.org/10.1016/j.procbio.2004.04.008

Aksu, Z., Açıkel, U., Kabasakal, E., & Tezer, S. (2002). Equilibrium modelling of individual and simultaneous biosorption of chromium(VI) and nickel(II) onto dried activated sludge. *Water Research*, *36*.

Al Ali, A., Ouda, M., Naddeo, V., Puig, S., & Hasan, S. W. (2021). Integrated electrochemical-adsorption process for the removal of trace heavy metals from wastewater. *Case Studies in Chemical and Environmental Engineering*, *4*, 100147. https://doi.org/10.1016/j.cscee.2021.100147

Bender, L. E., Lopes, S. T., Gomes, K. S., Devos, R. J. B., & Colla, L. M. (2022). Challenges in bioethanol production from food residues. *Bioresource Technology Reports*, *19*, 101171. https://doi.org/10.1016/j.biteb.2022.101171

Briffa, J., Sinagra, E., & Blundell, R. (2020). Heavy metal pollution in the environment and their toxicological effects on humans. *Heliyon*, *6*(9). https://doi.org/10.1016/j.heliyon.2020.e04691

Fu, F., & Wang, Q. (2011). Removal of heavy metal ions from wastewaters: A review. *Journal of Environmental Management*, *92*(3), 407–418. https://doi.org/10.1016/j.jenvman.2010.11.011

Ho, Y. S., & Mckay, G. (1999). Pseudo-second order model for sorption processes. *Process Biochemistry*, *34*.

Kanamarlapudi, S. L. R. K., Kumar Chintalpudi, V., & Muddada, S. (2018). Application of biosorption for removal of heavy metals from wastewater. In J. Derco & B. Vrana (Eds.), *Biosorption*. IntechOpen. https://doi.org/10.5772/intechopen.77315

Mazarji, M., Bayero, M. T., Sushkova, S., Mandzhieva, S., Tereshchenko, A., Timofeeva, A., Bauer, T., Burachevskaya, M., Kızılkaya, R., Gülser, C., & Keswani, C. (2021). Realizing United Nations sustainable development goals for greener remediation of heavy metals-contaminated soils by biochar: Emerging trends and future directions. *Sustainability*, *13*(24). https://doi.org/10.3390/su132413825

Mcdonald, C. W. (1977). Removal of toxic metal ions from metal-finishing wastewater by solvent extraction. *Separation Science*, *12*(4), 435–445. https://doi.org/10.1080/00372367708058088

Pohl, A. (2020). Removal of heavy metal ions from water and wastewaters by sulfur-containing precipitation agents. *Water, Air, & Soil Pollution*, *231*(10).

Qasem, N. A. A., Mohammed, R. H., & Lawal, D. U. (2021). Removal of heavy metal ions from wastewater: A comprehensive and critical review. *NPJ Clean Water*, *4*(1).

Redha, A. A. (2020). Removal of heavy metals from aqueous media by biosorption. *Arab Journal of Basic and Applied Sciences*, *27*(1), 183–193. https://doi.org/10.1080/25765299.2020.1756177

Scutarașu, E. C., & Trincă, L. C. (2023). Heavy metals in foods and beverages: Global situation, health risks and reduction methods. *Food*, *12*(18). https://doi.org/10.3390/foods12183340

Singha, S., Pasupuleti, S., Singha, S. S., & Kumar, S. (2020). Effectiveness of groundwater heavy metal pollution indices studies by deep-learning. *Journal of Contaminant Hydrology*, *235*, 103718. https://doi.org/10.1016/j.jconhyd.2020.103718

Sulaymon, A. (2014). Biosorption of heavy metals: A review. *Journal of Chemical Science and Technology*, *3*, 74–102.

Teow, Y. H., Sum, J. Y., Ho, K. C., & Mohammad, A. W. (2021). 3 - Principles of nanofiltration membrane processes. In N. Hilal, A. F. Ismail, M. Khayet, & D. Johnson (Eds.), *Osmosis engineering* (pp. 53–95). Elsevier. https://doi.org/10.1016/B978-0-12-821016-1.00014-0

Tezcan Un, U., & Öcal, S. (2015). Removal of heavy metals (Cd, Cu, Ni) by electrocoagulation. *International Journal of Environmental Science and Development*, *6*, 425–429. https://doi.org/10.7763/IJESD.2015.V6.630

Vijayaraghavan, K., & Yun, Y. S. (2008). Bacterial biosorbents and biosorption. *Biotechnology Advances*, *26*(3), 266–291. https://doi.org/10.1016/j.biotechadv.2008.02.002

Volesky, B., & Holan, Z. R. (1995). Biosorption of heavy metals. *Biotechnology progress*, 11(3), 235–250.

Wang, J., & Chen, C. (2009). Biosorbents for heavy metals removal and their future. *Biotechnology Advances*, *27*(2), 195–226. https://doi.org/10.1016/j.biotechadv.2008.11.002

Xiang, H., Min, X., Tang, C.-J., Sillanpää, M., & Zhao, F. (2022). Recent advances in membrane filtration for heavy metal removal from wastewater: A mini review. *Journal of Water Process Engineering*, *49*, 103023. https://doi.org/10.1016/j.jwpe.2022.103023

Yadav, M., Singh, G., & Jadeja, R. N. (2021). *Physical and chemical methods for heavy metal removal* (pp. 377–397). https://doi.org/10.1002/9781119693635.ch15

11 Scope and Challenges

Pavel Rout, Soma Maji, and Sumit Kumar

11.1 INTRODUCTION

Water pollution is a pressing global concern, driven by the discharge of industrial effluents, agricultural runoff, and domestic waste into our precious water bodies. These pollutants, ranging from heavy metals and synthetic dyes to emerging contaminants like pharmaceuticals, pose severe threats to both human health and ecological balance. Traditional wastewater treatment methods, while effective to a degree, often come with high processing costs, high energy demands, and the unfortunate consequence of generating secondary sludge that requires further management. Adsorption is a sustainable and efficient technology. This well-known physicochemical technique removes contaminants from water and air by accumulating chemicals (adsorbates) on a solid substrate (adsorbent). Adsorption is suitable for many environmental applications because of its simplicity, efficiency, and ability to regenerate and reuse adsorbents (Sen 2023).

Environmental water pollution arising from the significant release of many hazardous and harmful pollutants from human activities, including heightened industrialization, urbanization, population growth, and agricultural practices, poses a significant threat to human health and aquatic ecosystems. The prevalent heavy metal ions are Cu^{2+}, Ni^{2+}, Zn^{2+}, Cd^{2+}, Pb^{2+}, and Hg^{2+} ions. The World Health Organization (WHO) has recognized Cd^{2+}, Pb^{2+}, Hg^{2+}, and As^{3+} ions as the most hazardous heavy metal ions (Farhadi et al. 2021). Approximately 40% of Earth's surface water, predominantly consisting of river and lake water, is contaminated by heavy metal ions chiefly due to industrial and agricultural activities. The primary sources of heavy metal ion pollution include the discharge of untreated industrial effluents from refineries, coal-fired power plants, mining operations, alumina refineries, metallurgical industries, heavy chemical manufacturing, and chloro-alkali industries. Heavy metal ion pollution can lead to several detrimental health consequences, including diarrhea, gastrointestinal issues, paralysis, different dermatoses, hemoglobinuria, and vomiting (Kaçakgil and Çetintaş 2021). Heavy metal ions are very poisonous, detrimental to health, and non-biodegradable and provide a significant risk to the biosphere if not addressed. Consequently, there is an immediate necessity to provide environmentally sustainable and economically viable technology to eliminate these possible contaminants from the aqueous phase.

Several number of conventional techniques, such as chemical precipitation, oxidation, advanced oxidation, coagulation/flocculation, electrocoagulation, photocatalysis, membrane processes, reverse osmosis (RO), filtration, adsorption, solvent extraction, electroplating, ion exchange, activated sludge, and both aerobic and

 DOI: 10.1201/9781003767251-11

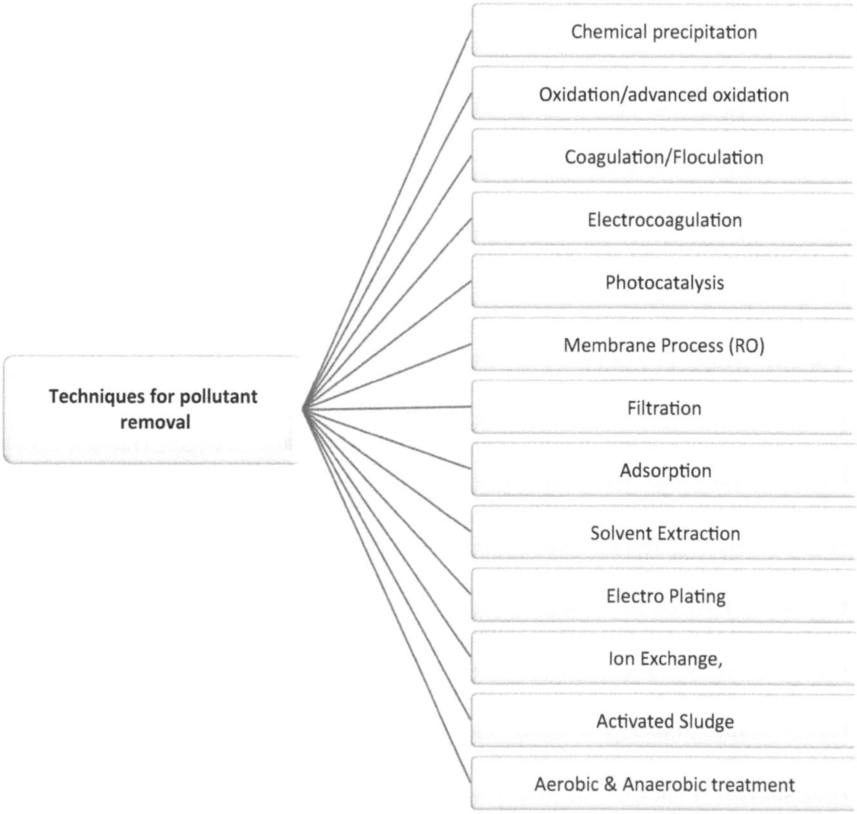

FIGURE 11.1 Conventional techniques to remove potential pollutants from wastewater.

anaerobic treatment, have been employed to eliminate these potential pollutants from water and wastewater, achieving varying degrees of efficacy (Figure 11.1). Adsorption-based separation technology is one of the most efficient and widely used technologies (Farhadi et al. 2021; Sabir et al. 2021). This is because it is easy to use, has a simple design, achieves high separation efficiency, works well at lower pollution amounts, is selective at the molecular level, uses little energy, and can separate multiple pollutant components with minimum secondary pollution (Bhatnagar and Sillanpää 2010).

11.2 SCOPES OF AGRO-INDUSTRIAL WASTES AS ADSORBENT MATERIALS

The "scope" of AIWs as adsorbent materials refers to their vast potential and the inherent advantages they offer, making them a highly promising and sustainable resource for environmental remediation. This section delves into the various facets that define this potential.

11.2.1 Availability of Sustainable Raw Materials

Different types of materials like rice husk, sugar cane waste, corn cobs, fruit peels, coffee grounds, sawdust, and other crop residues of essential human activities (agriculture and food processing) (depicted in Figure 11.2) ensure a sustainable and renewable supply (Pundlik et al. 2021; Sen 2023).

11.2.1.1 Corn Cobs

A by-product of corn harvesting, corn cobs are produced in large quantities globally, especially in countries with extensive corn cultivation (e.g., the United States, China, Brazil, and parts of India).

11.2.1.2 Wheat Straw

China is the number 1 wheat producer in the world. India is the second-largest wheat-producing country in the world. In the north Indian states (Punjab, Haryana, Uttar Pradesh, etc.), wheat is produced in a widespread manner. The availability of wheat straw is enough for the sustainable supply (Sen 2023).

11.2.1.3 Nut Shells

Groundnut shells, apricot shells, and other nut shells were available in India and other central and western Asian countries.

11.2.1.4 Sawdust

Woodworking, timber processing, and furniture manufacturing industries generate huge quantities of sawdust, woodchips, and bark, which are consistent and available waste streams (Pundlik et al. 2021).

11.2.1.5 Fruit Peels

Peels from citrus fruits (orange, lemon), banana, mango, and others are generated in large quantities from the industries of fruit processing, juice factories.

FIGURE 11.2 Several agricultural wastes that are used in the adsorption of metal ions.

11.2.1.6 Rice Husk

India, China, and other Southeast Asian countries produce millions of tons of rice husk. In West Bengal state of India, cultivation of rice carries out in a wide manner, which leads to enough supply of rice husk (Pundlik et al. 2021).

11.2.1.7 Sugarcane Bagasse

The residue of crushed sugarcane is immensely available in sugarcane-producing countries (Brazil, India, and Thailand). In India, Uttar Pradesh and Haryana are the major sugarcane producers (Sen 2023).

11.2.1.8 Other Residues as Coffee Grounds

The spent residue from coffee brewing is produced in huge volumes wherever coffee is consumed and processed, representing a substantial waste stream.

11.2.2 Inherent Physicochemical Properties for Adsorption

AIWs are naturally suitable as starting materials for adsorbent development, even before extensive modification (Simón et al. 2022).

11.2.2.1 Higher Concentration of Lignocellulosic Components

The primary components of most AIWs are cellulose, hemicellulose, and lignin (Natrayan et al. 2022; Pundlik et al. 2021).

- **Cellulose**: It contains hydroxyl groups (–OH), which are important for hydrogen bonding and help to bind different pollutants.
- **Hemicellulose**: A polysaccharide compound that also contains hydroxyl and acetyl groups, which are amorphous than cellulose and are easily degradable.
- **Lignin**: A phenolic polymer that provides rigidity and contains several functional groups (phenolic hydroxyls, methoxy groups), helping in metal binding and π–π interactions.

(Ensuncho-Muñoz and Carriazo 2015)

11.2.2.2 Presence of Natural Functional Groups

The surfaces of agricultural wastes contain various oxygen-containing functional groups, like:

- –OH groups mainly present in cellulose and hemicellulose and take part in metal chelation and hydrogen bonding (Natrayan et al. 2022).
- –COOH groups are abundant on the surface, work during ion exchange with cations, and increase the surface acidity.
- Phenolic groups originate from lignin and participate during reactions with metal ions for surface complexation (Pundlik et al. 2021).

- C=O groups are also present on the surface and contribute to the adsorption reactions.

These functional groups serve as active binding sites for various pollutants through mechanisms like ion exchange, surface complexation, and hydrogen bonding (Ensuncho-Muñoz and Carriazo 2015).

11.2.2.3 Porous Structure

Raw AIWs have natural fibrous and porous structures that make an open network. Although they may not have a large surface area in raw form, the use of physical and chemical stimulation can make this porosity a lot better. This will lead to larger surface areas and easier access for any molecules that hold pollutants (Bhatnagar and Sillanpää 2010).

11.2.2.4 Low Cost

The economic availability and low processing cost of AIW-based adsorbents are one of the major benefits.

- **Lower price of raw materials**: AIWs are generally recognized as waste products, with less economic value. For these reasons, those raw material cost was significantly lower than other conventional products (activated carbon products of coal or petroleum) (Bhatnagar and Sillanpää 2010).
- **Rural development**: Using local farm waste can help rural places grow their economies and make people's lives better by creating new businesses and jobs.
- **Waste management cost**: Industries and municipalities may substantially decrease their waste management and disposal costs (e.g., landfill tipping fees, incineration charges) by transforming wastes into a useful product (Demirbas 2008).

11.2.2.5 Environmental Sustainability

In addition to the economic advantages, using AIWs is also a great way to help the Earth. Diverting large amounts of AIWs in landfill as waste disposal alleviates the emission of methane (a powerful greenhouse gas), resulting from anaerobic decomposition. The circular economy model changes an economy from a "take-make-dispose" one to a circular one, where resources are used for as long as possible and waste is kept to a minimum. Adsorbents made from AIWs are less modified ones and contain a higher degree of biodegradability, which may be beneficial for eventual disposal, compared to persistent synthetic materials (Demirbas 2008).

11.2.2.6 Large Quantity of Pollutant Removal Capabilities

AIW-based adsorbents can successfully remove a wide range of contaminants from both water and air because they have a specific chemical composition and versatility in modification (Ensuncho-Muñoz and Carriazo 2015).

- **Heavy metals**: A strong affinity toward harmful metal ions, including lead, cadmium, chromium, copper, mercury, and arsenic. Some of the ways it works are ion exchange (H^+ for Mn^+), surface complexation (binding to –OH and –COOH), and chelation (Pundlik et al. 2021; Simón et al. 2022).
- **Organic dyes**: Effective removal of both cationic (e.g., methylene blue) and anionic dyes (e.g., Congo red) from textile and other industrial effluents. Mechanisms involve electrostatic interactions, hydrogen bonding, and π–π interactions between the dye molecule and the aromatic structures of the adsorbent (Demirbas 2008).
- **Pharmaceuticals and personal care products**: Promising for emerging contaminants like antibiotics (e.g., amoxicillin), analgesics (e.g., ibuprofen), hormones (e.g., estrogen), and personal care chemicals (e.g., triclosan). Adsorption often relies on hydrophobic interactions, pore filling, and electrostatic attraction/repulsion depending on the pH and pollutant charge (Simón et al. 2022).
- **Phenols**: Effective for phenols and derivatives from the petrochemical, pharmaceutical, and pulp and paper industries. Adsorption occurs via hydrogen bonding and hydrophobic interactions with the aromatic rings of lignin.
- **Pesticides**: Pesticides from agricultural waste water, often through hydrophobic interactions and specific chemical binding (Simón et al. 2022).
- **Gaseous pollutants**: Very efficient for entrapping CO_2, volatile organic compounds, such as benzene, toluene, and formaldehyde, sulfur oxides, and nitrogen oxides (Natrayan et al. 2022).

11.3 CHALLENGES IN AIW ADSORBENT DEVELOPMENT

Although AIWs show immense potential as adsorbents, their development and widespread utilization face several challenges. These challenges contain the inherent nature of the raw materials to technical, economic, and regulatory aspects, requiring concerted research and strategic planning to overcome (Sen 2023; Simón et al. 2022).

11.3.1 Raw Material Variability and Heterogeneity

One of the primary challenges stems from the diverse and often inconsistent nature of AIWs themselves. Unlike synthetic materials, AIWs are natural products susceptible to numerous influencing factors.

- **Seasonal availability**: In West Bengal, rice husk and jute sticks are primarily accessible following the harvest seasons of Kharif and Rabi rice and jute, respectively. This seasonal fluctuation can result in periods of scanty and oversupply (during harvest), which can complicate the commercial production of adsorbents on a year-round industrial scale (Sen 2023).
- **Geographical variation**: The chemical composition as well as functional qualities of the same variety of agro-waste can be significantly varying based on the soil type, climate, fertilizer use, and farming methods in the

area. For example, the amount of lignin in rice husk from different West Bengal areas (like Burdwan and South 24 Parganas) may be different, which could change the adsorbent's pore structure and functional groups. Because of this, it is hard to make an adsorbent that works the same way every time from raw materials that come from around the world or even from the same area (Sen 2023).

11.3.2 Pre-processing of AIWs

Raw AIWs require extensive pre-processing (Pundlik et al. 2021). This includes the following:

- **Collection and sorting**: Efficient and cost-effective collection and segregation of different types of agro-waste from dispersed agricultural fields or processing units.
- **Cutting and cleaning**: Cutting the materials into small pieces and cleaning from soil, dust, stones, unburnt carbon, or chemical residues (e.g., pesticides), which can contaminate the final adsorbent or interfere with the adsorption process.
- **Drying**: To prevent microbial decomposition, reduce transportation cost, and prepare the materials for further processing, drying is needed.

These pre-processing procedures contribute to the fact that the manufacture of adsorbents is more complicated and expensive overall.

11.3.3 Adsorbent Quality and Reproducibility

Production of a consistent and high-quality final adsorbent product is the major challenge (Simón et al. 2022). There is no standardized, widely accepted procedure for the pre-treatment, chemical/physical modification, and proper characterization of AIW-based adsorbents. There are only laboratory-scale methodologies for quality control. Even with carefully controlled laboratory conditions, achieving consistent adsorption capacities, selectivity, and regeneration efficiencies across different batches of AIW-derived adsorbents remains a significant challenge. Minor variations in temperature, concentration of activating agents, or reaction time can lead to substantial differences in the final product's performance, making scale-up problematic.

11.3.4 Limitations of Production and Scalability

Transforming a laboratory-scale synthesis into an economically viable and efficient large-scale production presents substantial engineering and logistical challenges.

- Pilot/industrial scale
  Standardization of different factors like reaction time, temperature control, mixing efficiency, and mass/heat transfer in larger vessels requires signifi-

cant knowledge and trials. For example, maintaining uniform temperature profiles during the pyrolysis of large quantities of rice husk is more challenging than in a small crucible (Sen 2023).

- Establishment costs
  Developing AIW-based adsorbents on an industrial scale requires substantial investment in specialized equipment (e.g., large-scale reactors, carbonizers, grinding mills, drying units, chemical storage, and handling systems) and the necessary infrastructure (e.g., waste collection points, processing facilities, and effluent treatment plants). This capital expenditure can be a barrier for new small and medium enterprises (Sen 2023).
- Higher energy requirements for activation
  Many effective modification techniques, particularly carbonization and pyrolysis, are energy-intensive processes (Farhadi et al. 2021). While the waste itself can sometimes serve as a fuel source, external energy input is often required. The high energy consumption can significantly offset the cost benefits of using free or low-cost raw materials, making the final product less economically competitive than traditional adsorbents.

11.3.5 Reusability Issues

For adsorption to be really sustainable and economically viable, the adsorbent must be regenerable and reusable throughout several cycles without substantial efficiency degradation. Adsorption capacity typically decreases after each regeneration cycle due to irreversible binding of some pollutants, damage to active sites, or structural changes in the adsorbent material. This limits the lifespan of the adsorbent and increases the overall operational cost. The regeneration methods (e.g., chemical washing with acids/bases, thermal desorption, biological degradation) are costly, are energy intensive, require significant amounts of reagents, or produce secondary waste streams that themselves need treatment (Natrayan et al. 2022). Repeated thermal or chemical regeneration cycles can lead to physical and chemical degradation of the adsorbent material. This includes pore collapse, loss of surface functional groups, or overall structural weakening, which ultimately reduces the adsorbent's performance and lifespan.

11.3.6 Competition with Existing Conventional Methods of Valorization

AIWs are also being used in different sectors nowadays. The demand for AIWs is increasing in other alternative uses. In West Bengal, rice husk is widely used as a fuel for boilers in rice mills and other small industries, for generating electricity, as a raw material for building materials (e.g., particle board), or for making silica. Sugarcane bagasse is a primary fuel source for sugar mill co-generation plants. Coffee grounds are used for composting or as a biofuel (Demirbas 2008). The economic attractiveness of developing and marketing AIW-based adsorbents must be competitive with these existing, often well-established, valorization pathways.

11.4 CONCLUSIONS

Environmental pollution, especially water pollution, is becoming a bigger problem, so it is essential to come up with new, long-lasting solutions now. Traditional ways of processing waste have some drawbacks, like being expensive, using plenty of energy, and generating additional waste. As a result, adsorption technology has become a very useful and flexible solution. However, it cannot be used more widely because common adsorbents like activated carbon are expensive and cannot be recycled. AIWs have become well-known because of the urgent need for long-term options. In places like West Bengal, India, a lot of agricultural and food processing waste is made every year, like rice husk and different fruit peels. This waste is not only a problem for getting rid of, but it can also be used to make high-performance adsorbent materials that do not cost much and last a long time. However, realizing this potential at scale faces significant hurdles, including the inherent variability of raw materials, challenges in ensuring consistent adsorbent quality, the complexities of scaling up production, and the critical issues surrounding regeneration and the safe disposal of spent materials. Overcoming these technical, economic, and logistical barriers demands concerted interdisciplinary research and robust collaboration between academia, industry, and policymakers. By addressing these challenges, AIW-based adsorbents are poised to become a cornerstone of sustainable environmental remediation, significantly contributing to a circular economy, minimizing waste burdens, and fostering a cleaner, healthier future globally and specifically in regions like West Bengal.

REFERENCES

Bhatnagar, A., and M. Sillanpää 2010. 'Utilization of Agro-Industrial and Municipal Waste Materials as Potential Adsorbents for Water Treatment—A Review'. *Chemical Engineering Journal* 157(2–3):277–96.

Demirbas, A. 2008. 'Heavy Metal Adsorption Onto Agro-Based Waste Materials: A Review'. *Journal of Hazardous Materials* 157(2–3):220–29.

Ensuncho-Muñoz, A. E., and J. G. Carriazo 2015. 'Characterization of the Carbonaceous Materials Obtained From Different Agro-Industrial Wastes'. *Environmental Technology* 36(5):547–55. doi:10.1080/09593330.2014.952679

Farhadi, A., A. Ameri, and S. Tamjidi 2021. 'Application of Agricultural Wastes as a Low-Cost Adsorbent for Removal of Heavy Metals and Dyes from Wastewater: A Review Study'. *Physical Chemistry Research* 9(2):211–26.

Kaçakgil, E. C., and S. Çetintaş 2021. 'Preparation and Characterization of a Novel Functionalized Agricultural Waste-Based Adsorbent for Cu2+ Removal: Evaluation of Adsorption Performance Using Response Surface Methodology'. *Sustainable Chemistry and Pharmacy* 22:100468.

Natrayan, L., S. Kaliappan, C. Naga Dheeraj Kumar Reddy, M. Karthick, N. S. Sivakumar, P. P. Patil, S. Sekar, and S. Thanappan 2022. 'Development and Characterization of Carbon-Based Adsorbents Derived From Agricultural Wastes and Their Effectiveness in Adsorption of Heavy Metals in Waste Water'. *Bioinorganic Chemistry and Applications* 2022(1):1659855. doi:10.1155/2022/1659855

Pundlik, R. C., S. D. Chowdhury, R. R. Dash, and P. Bhunia 2021. 'Life-Cycle Assessment of Agricultural Waste-Based and Biomass-Based Adsorbents'. *Biomass, Biofuels, Biochemicals* 22:669–95.

Sabir, A., F. Altaf, R. Batool, M. Shafiq, R. U. Khan, and K. I. Jacob 2021. 'Agricultural Waste Absorbents for Heavy Metal Removal'. In *Green Adsorbents to Remove Metals, Dyes and Boron from Polluted Water* (Vol. 49, pp. 195–228), Environmental Chemistry for a Sustainable World, edited by M. Inamuddin, I. Ahamed, E. Lichtfouse, and A. M. Asiri. Springer International Publishing.

Sen, T. K. 2023. 'Agricultural Solid Wastes Based Adsorbent Materials in the Remediation of Heavy Metal Ions From Water and Wastewater by Adsorption: A Review'. *Molecules* 28(14):5575.

Simón, D., C. Palet, A. Costas, and A. Cristóbal 2022. 'Agro-Industrial Waste as Potential Heavy Metal Adsorbents and Subsequent Safe Disposal of Spent Adsorbents'. *Water* 14(20):3298.

Index

A

Acetaminophen, 114
Acoustic cavitation, 119–160
Activated carbon adsorption, 60–123
Activated sludge limitations, 114–190
Adsorbate, 2–129
Adsorbent characterization, 4
Adsorbent interaction, 57
Adsorbent materials, 1–198
Adsorbent surface, 2–127
Adsorbent synthesis, 3–103
Adsorption capacity, 2–197
Adsorption efficiency, 3–181
Adsorption equilibrium, 43–46
Adsorption isotherms, 59–70
Adsorption kinetics, 4–179
Adsorption mechanisms, 2–84
Adsorption models, 4–72
Adsorption process, 2–196
Adsorption systems, 59–182
Advanced oxidation processes (AOPs), overview, 29–120
Agitation speed, 3–65
Agricultural industrial waste, 1–195
Agricultural residues, 1–101
Agro-industrial waste, 2–162
Air purification, 5–150
Algal, 101–102
Amino groups, 47
Anaerobic digestion, 22
Analytical techniques, 4–130
Anodic oxidation (AO), 118–170
Applications of adsorbents, 91–106
Aromatic compounds, 41
Aromatic ring interactions, 45–195
Atrazine, 5–48

B

Band gap, semiconductor, 123
Batch adsorption experiments, 4
BET surface area analysis, 4–139
Binding energy, 46–133
Bioaccumulation, 5–38
Bio aerogels, 10–21
Bio-based products, 8–18
Biochar, 45–150
Biochemical conversion, 9–14
Biodegradability enhancement, 17–194
Bioenergy, 9–151
Bioethanol production, 22
Biological modification, 174
Biomass combustion, 12
Biomass conversion, 9–14
Biomass energy, 10–22
Biomass feedstock, 14
Biomass fuel, 8–12
Biomass materials, 11
Biomass residues, 1
Biomaterials, 9–174
Bioplastics, 18
Biopolymer, 2–163
Biosorbent materials, 3, 178–180
Bulk catalysts, 123

C

Cadmium, 5–195
Calorific value, 9–12
Carbohydrates, 2–23
Carbonaceous adsorbents, 45
Carbonization, 4–197
Carbon nanomaterials (CNMs), 44–128
Carbon nanotubes, 45–145
Carbon sequestration, 23–151
Carboxyl groups, 102–177
Carcinogenic, 28–114
Catalysts, 42–134
Cellulose, 1–193
Characterization techniques, 4–144
Charge density, 46–48
Chemical activation, 4–162
Chemical adsorption, 2–127
Chemical characterization, 130
Chemical modification, 102–148
Chemical oxygen demand (COD), 120
Chemisorption, 2–184
Chlorinated hydrocarbons, 41
Circular economy, 83–198
Coagulation processes, 5–190
Coconut shells, 1–143
Combustion systems, 10
Composite materials, 80
Composting conditions, 18
Compression testing, mechanical properties, 134
Congo red dye, 173
Coniferyl alcohol, 14
Contaminant removal, 70–180
Contaminant removal efficiency, 79
Covalent organic frameworks (COFs), 45–104
Crop residues, 1–192
Crystalline, 4–144
Crystalline structure, 65–104
Crystallinity, 104–144

D

Decomposition, 36–196
Density, 9–173
Desorption process, 4–197
Diffusion, 4–185
    external mass transfer, 68–72
    pore diffusion mechanism, 68
    surface diffusion, 71–160
Diffusion models, 4
Dipole–dipole interactions, 48–66
Disinfection, 54–55
Disintegration properties, 18
Dissociation constant (pKa), 44
Donor–acceptor, 45–48
Drying methods, 3–197
Dye adsorption, 49–72
Dye removal, 2–106
Dyes
    cationic dyes, 67

E

Electrochemical AOPs (EAOPs), 115–171
Electron-hole pair formation, photocatalysis, 123
Electrostatic attraction, 44–195
Electrostatic interactions, 2–195
Electrostatic repulsion, 44–169
Emerging contaminants, 41–195
Emerging pollutants, 5–110
Energy changes (Gibbs free energy), 4–97
Energy density, 11
Energy recovery processes, 150–152
Engineered nanomaterials, 41
Enthalpy change, 58
Entropy, 4–97
Environmental contamination, 15–167
Environmental pollution, wastewater, 1–198
Environmental remediation, 1–198
Environmental sustainability, 77–194
Enzymatic hydrolysis, 15–22
Equilibrium isotherms, 4
Ethanol, 9–174

F

Factors affecting adsorption
    adsorbent properties, 64–75
    pH influence, 177
Feedstock, 6–152
Fermentation, 3–157
Fermentation processes, 22
Fiber, 2–161
Filtration systems, 5–109
Floriculture, 18–19
Fourier-transform infrared spectroscopy (FTIR), 4–137
Freundlich isotherm, 58–182
    heterogeneous adsorption, 61
Frontier orbital theory (FOT), 46
Fuel, 8–197
Functional group, 2–197
Functionalization techniques, 4

G

Gaseous pollutants, 5–195
Gasification process, 9–22
Gibbs free energy (ΔG), 4–97
Global energy demand, 22
Glucopyranose units, 15
Graphene-based adsorbents, 45–153
Green biotechnology, 10
Greenhouse gas, 1–194
Grinding processes, 3–197
Groundwater, 5–186

H

Heating value, 9–151
Heavy metal adsorption, 81
Heavy metal adsorption mechanisms, 81
Heavy metals
    adsorption removal, 80–81
    lead, cadmium, arsenic, 5–197
Hemicellulose, 1–193
Heteropolysaccharides, 16
Hydrogel formation, 21–47
Hydrogen bonding, 3–195
Hydrogen bonding interactions, 48
Hydrophobic interactions, 44–195
Hydroxyl groups, 47–193

I

Impregnation, 3–4
Industrial effluents, 5–195
Industrial waste, 1–197
Insulation materials, 18
Intra-particle diffusion model, 184–185
Ion exchange, 3–195
Ionic strength influence, 43–187
Isotherms
    BET model, 4–144
    Freundlich model, 61–185
    Langmuir model, 63–185

J

Juice processing waste, 107–192

K

Kinetic models, 4–184
Kinetics of adsorption, 71–172
  pseudo-first-order model, 4–184

L

Landfill waste biomass, 11–194
Langmuir isotherm, 58–182
Lead, 5–197
Lignin, 1–196
Lignocellulosic biomass, 1–21
Lignocellulosic feedstock, 150
Low bulk density fuels, 12

M

Macro-elements in ash, 12
Magnetic porous carbon sorbents (MPCS), 44
Mass transfer
  external diffusion, 68–72
  rate determining step, 129–156
Mechanical strength, 14–187
Membrane filtration, 5–168
Metal ion adsorption, 2–74
Metal-organic frameworks (MOFs), 2–195
Microbial, 3–196
Microbial contaminants, 5
Micro-elements in biomass ash, 12
Microfibril structure, 16
Microplastics, 5–110
Microplastics adsorption, 5–110
Mineralization, 115–120
Modification techniques, 3–197
  activation, 3–197
  functionalization, 3–159
Moisture, 18–151
Molecular electrostatic attraction, 44
Molecular interactions, 61–84
Municipal biomass utilization, 8–30

N

Nanocellulose materials, 21
Nanoparticle, 4–154
Nanoparticle adsorption systems, 4–154
Non-covalent interactions, 46
n–π interactions, 45–49
Nutrient, 32–173

O

Organic matter composition, 17–54
Organic micropollutants (OMPs), 41–44
Organic pollutant, 2–167
Oxidative stress, 20–37
Oxidizing agents, 3–115
Ozonation processes, 27–118
Ozone (O_3), oxidation potential, 54–129

P

Packed bed reactors, 5
Parameter, 61–186
Pentose sugars, 41–21
Persistent organic pollutants (POPs), 42–118
Pesticide removal, 5–196
pH, 3–195
Pharmaceuticals, 5–195
pH buffering range, 46
Phenolic compounds, 2–21
pH influence, 5, 177
Photocatalysis, 27–191
Photochemical AOPs, 115
Photolysis, mechanism, 121–122
Photosynthesis, 10–121
Physicochemical properties, 4–193
Physisorption, 2–92
  van der Waals forces, 2–133
π–π interactions, 3–195
Plant biomass, 15–99
Pollutant, 6–194
Polycyclic aromatic hydrocarbons (PAHs), 36–152
Polymer chain, 15–193
Polymeric adsorbents, 78–82
Pore size distribution, 4–169
Pore size distribution, analysis, 4–169
Porosity, 4–194
Precipitation, 5–171
Pretreatment processes, 15–181
Processing by-products, 1
Proteins, 2–174
Protonation, 44
Pseudo-first-order model, 4–184
Pseudo-second-order model, 4–184
PXRD (powder X-ray diffraction), 131
Pyrolysis, 4–197
Pyrolysis characteristics, 11
Pyrolysis temperature, 150

Q

Quality, 5–198
Quality assessment, 33

R

Reaction rate theory, 71–75
Regeneration of adsorbents, 56–198
Regulatory considerations, 6
Removal efficiency (R%), 43–180
Renewable feedstock, 8

Renewable resources, 15–158
Resins, 161–162
Resource recovery, 5–77
Reverse osmosis, 42–190
Rice husk adsorption, 1–198

S

Sawdust adsorbents, 1–192
Scalability, 6–196
Scalability issues, 6–155
Scanning electron microscopy (SEM), 4–142
Selectivity, 42–196
Semiconductor, 123
Sludge reduction, 118
Sonolysis, 119–120
Sulfate radicals, 115
Surface area, 3–194
Surface charge properties, 3–171
Surface chemistry, 3–141
Surface complexation, 3–195
Surface functional groups, 3–197
Surface functionalization, 58–153
Surface interactions, 55–81
Surface morphology, 4–142
Surface properties, 3–154
Surfactants adsorption, 41
Sustainability, 5–194
Sustainable water treatment, 64
Syngas production, 150
Synthetic dyes, 3–190

T

Temkin isotherm, 4
Temperature effects, 5
Textile, 5–195
Thermal conductivity, 18
Thermal treatment, 3–148
Thermochemical, 5–14
Thermo-chemical conversion, 11–14
Thermodynamic parameters, 4
Tomato waste adsorbent, 3
Toxicity reduction, AOPs, 20–171
Transmission electron microscopy (TEM), 4–144

U

Ultrafiltration, 169
Ultrasound irradiation, 119
Ultrasound treatment, 115–161
Utilization of waste, 1

V

Van der Waals forces, 2–133
Volatile organic compounds (VOCs), 5–195

W

Wastewater treatment, 5–190
Water absorption properties, 135
Water purification, 38–170
Water treatment systems, 62–107
Wheat bran adsorbent, 2
Wheat straw, 1–192
Wood, 1–150
Woody biomass, 22

X

X-ray diffraction (XRD), 4–144
X-ray fluorescence (XRF), 132
X-ray photoelectron spectroscopy (XPS), 40–134

Y

Yield optimization, 11–120

Z

Zeolites, 55–161
    ion exchange capacity, 67–73

For Product Safety Concerns and Information please contact our
EU representative GPSR@taylorandfrancis.com Taylor & Francis
Verlag GmbH, Kaufingerstraße 24, 80331 München, Germany